Map of
SOUTH AFRICA,
made in 1856
at the time of Livingstone's
journeys, shewing the amount
of Geographical knowledge
then available.

50   100   200   300
English Miles.

The Rev.ᵈ Dr Livingstone's Route is Coloured

DAVID LIVINGSTONE
Letters & Documents
1841·1872

DAVID LIVINGSTONE

Letters & Documents
1841·1872

**The Zambian Collection
at The Livingstone Museum**

containing a wealth of
restored, previously unknown
or unpublished texts

**Edited by
TIMOTHY HOLMES**

with Chronology,
Biographical Chapters and
Annotated Index

THE LIVINGSTONE MUSEUM
LIVINGSTONE
in association with

MULTIMEDIA ZAMBIA
LUSAKA

INDIANA UNIVERSITY PRESS
BLOOMINGTON AND INDIANAPOLIS

JAMES CURREY
LONDON

Published on behalf of
The Livingstone Museum, Livingstone

By James Currey Ltd
54b Thornhill Square, Islington, London N1 1BE

Indiana University Press
10th and Morton Streets, Bloomington, Indiana 47405

Multimedia Zambia
P.O. Box 320199, Lusaka

First published 1990

British Library Cataloguing in Publication Data
Livingstone, David, *1813—1873*
Letters and documents 1841—1872.
1. Africa. South of the Sahara. Exploration Livingstone, David, 1813—1873
I. Title II. Holmes, T.
916.7′04

ISBN 0—85255—041—3

Library of Congress Cataloging-in-Publication Data
Livingstone Museum.
David Livingstone: letters & documents, 1841—1872: the Zambian collection at the Livingstone
Museum, containing a wealth of restored, previously unknown or unpublished texts. Edited by
T. Holmes.
p.   cm.
ISBN 0—253—33516—7 (Indiana University Press)
1. Livingstone, David, 1813—1873—Correspondence.
2. Explorers—Africa—Correspondence.   3. Explorers—Scotland—Correspondence.
I. Livingstone, David, 1813—1873.   II. Holmes, Timothy.   III. Title.
DT3081.L58A4   1990
916.89404′1—dc20   89—24725
                                                            CIP

Set in 10/12pt Bodoni by Colset Private Limited, Singapore
and printed by Villiers Publications London N6

In memory of
Dr Kafungulwa Mubitana
1938–1978
The first Zambian Director of
The Livingstone Museum

Kafungulwa Mubitana
was born at Namwala, Zambia on 16 October 1938.
He was educated at Munali Secondary School, Lusaka
at Makerere University, Uganda and at Edinburgh University, Scotland.
He was awarded a Doctorate of Philosophy (Social Science) in 1977.
He became Director of the Livingstone Museum in January 1973
and died in a motor accident on 25 June 1978.

CONTENTS

The four-figure numbers refer to G W Clendennen and
I C Cunningham, *David Livingstone: A Catalogue of Documents*
(Edinburgh 1979/1985) and the names that follow to the relevant editors
or biographers, with (*i*) if incomplete, and (*e*) if extract only, on previous
publication.

PART ONE

The Early Years

Contents

PART TWO
The Zambezi Expedition

Contents

PART THREE
Interlude in Britain

Contents

Contents

PART FOUR
The Last Journey

(Based on the map in G. Seaver *David Livingstone, His Life & Letters* (1957) and reproduced by kind permission of the Lutterworth Press)

FOREWORD

In writing a foreword to the publication of these letters and documents of Dr David Livingstone, I want personally to welcome all our readers' interest in this book — the first published by the Livingstone Museum.

The museum, which is the largest and oldest in Zambia, is situated in the centre of Livingstone, the tourist capital of Zambia, and a few kilometres from the Victoria Falls on the Zambezi River. The museum is an impressive structure and forms a depository for Zambia's cultural, historical and natural heritage.

The history of the museum dates back to 1930, when colonial administrators began to purchase ethnological and ethnographic materials from all over the country. This was a result of the awareness of the significance of preserving the material culture of the Zambian people. In 1934 the idea was conceived of extending the collections of traditional arts and crafts to become a memorial to Dr David Livingstone, and to open the collection to the public. Sir Hubert Young, the then colonial governor, collected several letters written by David Livingstone, and the collection was extended over the years by the addition of donations from Livingstone's descendants, from descendants of his correspondents, and by purchase; besides the manuscripts, the museum has also been the recipient of numerous Livingstone relics as well as a number of note- and sketchbooks. Some of these precious items are on permanent display in the history gallery, while the rest are stored in the museum's research library.

In 1934, the museum's first exhibition, which included much Livingstone material, was opened to the general public in the Old Magistrates' Court, Livingstone, awaiting the building of the present structure.

This was originally named the Rhodes—Livingstone Museum, but was changed to the simple Livingstone Museum in 1966. It now has well over 12,000 specimens of different objects of Zambia's traditional material culture, over 5,000 historical items and some 20,000 natural history specimens, and is considered one of the best museums in independent Africa, with its sequences showing the development of mankind in Zambia from the early stone age to recent decades.

The museum has four specialist departments, history, prehistory, ethnography and art, and natural history. It also has four permanent galleries, a temporary exhibition gallery, and a craft shop.

Our main purpose in publishing David Livingstone's letters is to make them available to Zambian readers, hence the detailed index. It is hoped that this publication will further enlighten many readers as regards the place of Livingstone in Zambia's history, and that of neighbouring countries. Furthermore, I hope that

Foreword

it will be possible for many of our readers to learn and understand better David Livingstone's life as a missionary, geographer, and explorer.

Reading through his letters, I find Livingstone's life to be an incredible one which could only have been lived by an extraordinary human being — and to me he was indeed an extraordinary person.

MWIMANJI CHELLAH
Director
The Livingstone Museum
Zambia
30 November 1989

INTRODUCTION

'Well, to look at the man you would think nothing of him; but he is a plucky little devil.'[1]

The one hundred and forty letters, fragments and other pieces in this collection are arranged in the order in which they were written, and are presented in four parts: The Early Years, 1841–54; The Zambezi Expedition, 1858–64; Interlude in Britain, 1864–5; and The Last Journey 1865–72. Each part is introduced with a chronology of Livingstone's life over the period and a comment on his career, forming together a brief biography. The appendices provide a further perspective on his interests, while background notes are incorporated in the index.

The texts span the greater part of Livingstone's life as missionary, geographer and explorer, but they form only a small part of a total correspondence which amounts to over two thousand letters, besides numerous despatches to his employers and sponsors.

All of Livingstone's writings known up to 1985 are listed in G W Clendennen and I C Cunningham's *David Livingstone: A Catalogue of Documents* (1979) and its Supplement (1985), published by the National Library of Scotland, Edinburgh, for the David Livingstone Documentation Project: but that twenty pieces in this Zambian collection are not recorded in the Catalogue raises the hope that still more material may come to light in private holdings or in libraries and archives.

It is the fate of persons who make an impact on their times to have their lives investigated thoroughly and perhaps impertinently by those who come after them. David Livingstone is no exception to this rule, nor has he suffered less than other historic figures in the interpretations put on his life's work by biographers, historians, and even some editors of his Letters and Journals — the latter by deliberate suppression.

An example of this practice is to be found in D Chamberlin *Some Letters of David Livingstone 1840–72* (London 1940), where the editor, despite his disclaimers, has made substantial cuts to the texts. G. Seaver in *David Livingstone: His Life and Letters* (London 1957) is inclined to do the same, as is W E Oswell in *William Cotton Oswell* (London 1900), the biography of his father, Livingstone's friend of many years' standing.

It is regrettable that these editors did not have the scholarly principles of Professor I Schapera to guide them. His editions of Livingstone's *Family Letters*

---

[1] C J Andersson *The Okavango River* (London 1861), p 84: the speaker was probably William Cotton Oswell.

(London 1959), *Missionary Correspondence* (London 1961), *South African Papers* (Cape Town 1974), as well as the *Private Journal* (London 1960) and *African Journal* (London 1963), are models of thoroughness and objectivity which many will strain to emulate. Admirable followers of Schapera's example have been R Foskett *The Zambezi Doctors — David Livingstone's Letters to John Kirk 1858–72* (Edinburgh 1964) and M Boucher *Livingstone Letters 1843–1872* (Johannesburg 1985). See also G. Shepperson, *David Livingstone and the Rovuma: a Notebook* (Edinburgh 1965).

The letters in the above volumes make up about half of the total, while the remaining unpublished Journal (for 1861–3) has been prepared for print by G W Clendennen. *The Last Journals of David Livingstone from 1865 to his Death* (London 1874) was edited, with excisions, by H Waller; J P R Wallis's *The Zambezi Expedition of David Livingstone* (London 1956), an edition of the 1858–61 Journals, is likewise marred by cuts.

Taken with Livingstone's *Missionary Travels and Researches in South Africa* (London 1857) and *Narrative of an Expedition to the Zambezi and its Tributaries* (London 1865), most of his writings have appeared in print (which is not to say that they have stayed in print): however, as we have seen above, many letters remain unpublished despite a welcome flow in academic journals.

Biographies of Livingstone are legion, but none gives a satisfactory account of his life or character. Three of the most substantial are W G Blaikie *The Personal Life of David Livingstone* (London 1880), G Seaver (see above) and T Jeal *Livingstone* (London 1973). Livingstone as geographer is examined in F Debenham *The Way to Ilala* (London 1955) and as physician in M Gelfand *Livingstone the Doctor* (Oxford 1957); specific periods of his career are covered in R Coupland *Livingstone's Last Journey* (Oxford 1928) and G Martelli *Livingstone's River: a History of the Zambezi Expedition* (London 1970), both thorough and well documented, though Coupland's conclusions may be open to question. Questionable too is Blaikie's saintly tone, which also infects Seaver, while Jeal, like J Listowel *The Other Livingstone* (Lewes 1974), tends to the opposite extreme. O Ransford *Livingstone The Dark Interior* (London 1976) attempts to prove that the author's subject was a congenital manic-depressive.

Apart from these works, there exists a host of popular biographies, such as R W Dawson *Livingstone, the Hero of Africa* (London 1923) and S S Starritt *Livingstone the Pioneer* (London 1927?), not to mention plagiarisms such as that perpetrated by Adama (see Index).

A balanced assessment of a man whose life was even by the coolest reckoning a most extraordinary one has yet to be written, while the *Collected Letters of David Livingstone* remains, unfortunately, no more than a distant prospect.

Only ten of the texts in the Zambian collection have been published before in full, and although much of the ground the edition covers is already known from Livingstone's Journals, or other parts of his correspondence — he often wrote similar letters simultaneously to a number of persons — new aspects of his life and character may emerge in this volume.

The letters to James Young show Livingstone's interest in business and suggest

Introduction

that he planned to become a trader himself. His polemic against the Portuguese statesman Sá da Bandeira hints that Livingstone in 1862 anticipated the 1880s partition of Africa among the colonial powers. The correspondence with W C Oswell in 1864–5 gives a clearer idea than hitherto of the way in which *Narrative of an Expedition* . . . was produced, and of Livingstone's attitude to Victorian social mores. The 1866 letter to George Frere illustrates Livingstone's scorn for British hypocrisy in east African slave-trade policy, while the account to Sir Morton Peto of Bedingfeld and the ship *Bann* may shed new light on an old quarrel. Several letters inject fresh venom into the Baines controversy. All in all, the reader of this Zambian collection is left with a clearer idea of Livingstone's wide range of interests, of the breadth of his reading, of the extent of his social circle, of his ambitions, and of his sometimes scatological sense of humour.

Two frequent motifs in the letters are Livingstone's dedication to Christ and his determination to see an end to slavery. An understanding of his approach to these matters is essential to any appreciation of his life, but it is impossible to examine them in any detail here, though a few words may be said:

Religion: Livingstone belonged to a tradition of Christianity, elaborated by the French reformer Jean Calvin and his Scottish follower John Knox, which adheres to the doctrine of predestination: God knows at the time of conception which of his human creatures is destined for heaven and which for hell. In life not even a believing and practising Christian can be sure of salvation, while all non-Christians are damned no matter how virtuous they may be. A believer, however, can obtain some assurance of salvation by being consciously filled with trust in Christ: apart from this consolation, a further sign of salvation is the achievement of material wealth and success — a sign of God's blessing. The love for and trust in Christ which Livingstone felt throughout his adult life may possibly have assured him that he was one of the chosen, but the apparent lack of success which marked his last years could have caused serious concern about his ultimate destination.

Slavery: Human bondage is as old as history itself, and was approved for many centuries in Christianity. However, during the 18th century, and contemporary with the rise of industrial civilization in Europe, a revulsion against slavery began to develop there, additional to the revulsion of the slaves themselves which was expressed in innumerable rebellions in the colonies. In 1833 Britain banned slavery throughout its empire, having outlawed the trade in slaves some twenty years previously. The Acts of Parliament which gave these measures the force of law followed long abolitionist campaigns waged by former slaves, members of the new industrial working class, philanthropists, industrialists and clergymen — campaigns which emphasized the inhumanity of slavery and its incompatibility with true Christian moral teaching. The British laws were strengthened by treaties with other colonial powers such as Portugal, and were enforced by patrols at sea, first mainly in the Atlantic, but in Livingstone's time and after in the Indian Ocean as well.

Besides the abolitionists' undoubted humanitarian motives, they were supported by sound, long-term, economic considerations. Industrial civilization was based on new technology and on the production and export of factory-made goods. Such a market the slave states could never become, but it was equally clear that if slavery

Introduction

were replaced by wage labour a wide outlet for manufactures would come into existence, while abolition would also lead to the replacement of much manual labour by industrial machinery imported into the plantation economies.

All these factors would benefit industry by strengthening its base and by offering its own exploited underclass at home the promise of enough prosperity to dissuade it from revolution. A further consideration was that abolition could pre-empt the constant threat in slave colonies of any repetition of the 1804 revolution in Haiti which had turned that French possession into an independent state. It was also foreseen that with the end of slavery the lands of tropical Africa from which the majority of slaves were exported would retain a sufficient workforce to enable them to become producers of the agricultural raw materials which industry needed but could not grow in its cold northern heartlands.

Livingstone was born into that heartland, into a part of society which would benefit directly from increasing industrial production. Whether, in his mind, the humanitarian or the economic motive were uppermost, he was a natural abolitionist.

This did not restrain him from enjoying the hospitality of slave-owners in Angola, Moçambique and Zanzibar, or of slave merchants in the African interior — he was pragmatic enough to understand that slavery would only die when it was replaced by industrial civilization — which he believed to be the apotheosis of Christianity. He pursued this objective to the end of his days.

Editorial method

All the David Livingstone texts and illustrations in this edition are transcribed or copied from material in the Zambian collection. With the exception of letters 22, 24 and 125 the transcription has been made from holograph; the three exceptions are from typescripts of the originals. In transcribing, Livingstone's spelling has been left as it stands, but the punctuation, which is erratic, has been slightly modified. Dubious readings and uncertain dates are placed between square brackets. Illegible words are indicated by a question mark between square brackets. Italics have been used in place of Livingstone's underlining.

Sources of annotation

Much of the information would have been impossible to obtain without reference to *David Livingstone: A Catalogue of Documents* and its supplement, and to the various editions of Letters and Journals mentioned above, or without access to the resources of the National Archives of Zambia, Lusaka; the Bodleian Library, and especially the Rhodes House Library, Oxford; the National Library of Scotland, Edinburgh; the Arquivo Histórico Ultramarino, Lisbon; the Archives of Strathclyde University, Glasgow; the Scottish National Memorial to David

Introduction

Livingstone, Blantyre; and the Killie Campbell Africana Library, University of Natal, Durban. To the persons at these institutions who helped me, the warmest thanks are extended.

Indispensable in preparing this edition have been the (British) *Dictionary of National Biography*, and its Netherlands, French, Portuguese, United States of America, and South African counterparts, as well as *Chambers Biographical Dictionary*.

ACKNOWLEDGEMENTS

The editor was awarded the Richard Hillary Prize, 1986 (administered by Trinity College, Oxford) to enable him to travel from Zambia to Europe to pursue his research. The editor was also elected Temporary Member of Common Room, Wolfson College, Oxford for the same purpose. An inestimable debt of gratitude is owed to the Richard Hillary Trustees and to the President and Fellows of Wolfson College, for without their support completion of this edition would have been impossible.

Many individuals have given assistance in one way or another, and those who have made direct contributions are acknowledged after the appropriate entry. Special thanks are due to the Director of the Livingstone Museum, Zambia, Mr M Chellah; to the Museum's Keeper of History, Mr S Mudenda, to the Archivist, Mr M Michello and the Technical Officer, Mr R Lupiya; and also to Mrs N E Holmes (Lusaka), Ms M Sifuniso (Lusaka), Professor L Nkosi (Lusaka), Dr S Crehan (Lusaka), Mr and Mrs C Laugery (Lusaka), Mr E Chuma (Lusaka), Mr and Mrs M G B Pilcher (Lusaka) and Dr and Mrs M Bush. In Britain, invaluable help and support came from Ms K Brown, Mr and Mrs J R Vigne, Mr and Mrs D H Burden, Mr and Mrs P Katjavivi, Sir Raymond and Lady Hoffenberg, Mr and Mrs G E M de Ste Croix, Mr I C Cunningham, Dr J Macgrath, Mr W Cunningham. A friendly letter from Mrs D Harryhausen provided encouragement to proceed with the project.

The hospitality and understanding of Mr and Mrs J Currey have been exceptional, most notably in the generous offer by James Currey Limited of the typeset film of the book to facilitate parallel publication in Zambia by the Livingstone Museum, at no financial gain. The Livingstone Museum, Multimedia Zambia and James Currey Publishers acknowledge with deep appreciation the contributions of the Pollock Memorial Missionary Trust and the Richard Hillary Trust of Trinity College, Oxford towards the Zambian edition of this book.

Timothy Holmes
Lusaka

PART ONE
THE EARLY YEARS
1813 • 1858

CHRONOLOGY

1813	19 March	Born Blantyre, Scotland
1823		Started work in cotton mill
1836		Entered Anderson's College, Glasgow
1839		Probationer, London Missionary Society
1840		Graduated Doctor of Medicine; member of LMS; ordained; sailed for south Africa
1841	July	Arrived at Kuruman
1843		Founded mission at Mabotsa
1845	2 January	Married Mary Moffat
1846		Attempted to found mission at Chonwane; Robert born
1847		Founded mission and settled at Kolobeng; Agnes born
1849		Visited Lake Ngami with WC Oswell; Thomas born
1850		Tried but failed to visit Sebitwane; met FW Webb; Elizabeth born but died an infant
1851		Visited Sebitwane; William Oswell born; sent Mary and children to Britain
1852		Kolobeng mission ransacked by Boers
1853		Return to Sebitwane's country; Sebitwane dead but successor Sekeletu supports Livingstone's plans
1854		Completed journey to Luanda, Angola, return to Sekeletu; journey to east coast
1855		Named Mosi-oa-Tunya 'Victoria Falls'
1856		Arrived at Quelimane, Moçambique;
1856	December	London

| 1857 | Published *Missionary Travels and Researches in South Africa*; speeches at Oxford and Cambridge which led to formation of Universities' Mission to Central Africa |
| 1858 | Resigned from LMS; appointed British Consul to 'Quelimane and to Rulers in the Interior', and leader of Expedition to the Zambezi |

David Livingstone was born in 1813, the second of five surviving children of Neil and Mary Livingstone, at Blantyre, near Glasgow, Scotland, then one of the principal industrial cities of Britain. His father earned a living as a peddler of tea, and was also a propagandist for an independent congregation which had broken away from the Church of Scotland and established its own place of worship at Hamilton. Neil imposed strict religious discipline on his family, but also helped his children to get a basic education. At the age of ten, David had to go to work in the Blantyre cotton mill (in one of whose lodgings the family lived) and there he remained for thirteen years, when his father agreed to his becoming a medical missionary. David had continued his education while at the mill, and had also taught himself Latin, giving himself sufficient qualifications to be enrolled at Anderson's College, Glasgow, in 1836, at the age of twenty-three.

Three years later he was accepted on probation by the London Missionary Society, and moved to the LMS training institution at Ongar, near London: at the same time he completed his medical studies in London, and graduated as doctor of medicine in 1840, the year he was ordained after being admitted to full membership of the LMS. At first he had hoped to work in China, but when that proved impossible, he accepted a posting to Kuruman in south Africa, where Robert Moffat, whom he had met in London, had established a mission in 1821. Livingstone travelled by sea to Cape Town, where he met the local LMS officials, on to Algoa Bay (Port Elizabeth), and then overland by wagon to the LMS station at the southern edge of the Kalahari.

At the time of Livingstone's arrival in 1841, south Africa was in turmoil. The British had taken over the Cape Colony from Holland in 1806, and in 1820 brought in new settlers to occupy the eastern frontier areas, where there was a continuous state of war as Africans resisted European encroachment. Moreover, the abolition of slavery in the now-British colony in 1834 had led a large number of Dutch-speaking settlers to trek north out of the colony in protest and establish themselves in what are today the Orange Free State, the Transvaal and Natal.

In 1841, Kuruman still lay outside the Cape Colony, in independent Tswana lands, but all along the eastern fringes of the Kalahari there was friction verging on war between the Batswana and the Boers.

A further manifestation of the general turmoil had been the growth of the Zulu empire during the early years of the century, when Shaka in particular had attempted to build a state strong enough to resist European expansion. Peoples in or around Shaka's territory who would not submit to Zulu rule were driven out or moved out (the mfecane*), among them Mzilikazi's Ndebele, who settled first in the Transvaal, then went north to Zimbabwe after being defeated by the Boers. Other important migrants were the Kololo, led by Sebitwane from the Fokeng territory in today's Orange Free State. The Kololo first moved south-west, but on being stopped by the Griquas near Kuruman, skirted the Kalahari and eventually settled along the Upper Zambezi (known locally as the Lyambai) where they subjugated the Lui (Lozi), an offshoot of the Lunda empire to the north. Sebitwane's sway extended as far as the Batoka plateau, which made the Kololo neighbours of the*

3

Ndebele. There was continuous enmity between the two states across Zambezi.

Livingstone began to travel north from Kuruman through Tswana territory shortly after his arrival, partly to improve his knowledge of the language, Setswana, and partly to find a site for a mission of his own. After false starts at Mabotsa and Chonuane, he finally settled at Kolobeng, near the town of Chief Sechele of the Bakwena. Livingstone had meanwhile married Moffat's daughter Mary, and they had started a family, five children being born during the period — Robert Moffat (1846), Agnes (1847), Thomas Steele (1849), Elizabeth (1850), who died as a baby, and William Oswell (1851).

He had also met and made friends with several wealthy Englishmen who came to the country to hunt elephants and other big game. Among them were Thomas Steele, William Frederick Webb, Frank Vardon, and William Cotton Oswell: two of the Livingstones' sons were named after Steele and Oswell.

In 1851 Livingstone and W C Oswell travelled around the northern Kalahari to Lake Ngami, the first Europeans to do so. Livingstone's report of the expedition and its successful outcome, to the LMS headquarters in London, brought recognition and a prize from the Royal Geographical Society. A later journey with Oswell took Livingstone to the Chobe and upper Zambezi rivers, where they met Sebitwane. Livingstone was pleased to find that the Kololo language was similar to Setswana (which he now spoke fluently) and that Sebitwane wanted him to establish a mission in his country. The king may or may not have been interested in Christianity, but whatever the case, a prime element of Kololo foreign policy was relations with Mzilikazi and the Ndebele, who were regarded as a threat. Moffat of Kuruman was on good terms with the Ndebele king, and it was thus possible that the permanent presence of Moffat's daughter and son-in-law among the Kololo would still any aggressive intentions that the Ndebele might harbour.

By the time of his meeting with Sebitwane, Livingstone was disillusioned with orthodox missionary work (he had made only one convert in twelve years) and was questioning the policies of the LMS. He was also beginning to develop the thesis that 'Christianity, Commerce, and Civilization' were inseparable, and came to believe that the Kololo kingdom, with its apparently abundant resources, so unlike the Kalahari domain of the Batswana, was a suitable area to put it into practice. However, the route from the coast at the Cape was long and dangerous, both because of the desert and because of Boer hostility to any British advance flanking the Transvaal. Livingstone decided to find an alternative.

He took his wife and children to Cape Town, where at Oswell's expense — with money raised from the sale of ivory — they were kitted out and sent to Britain, there, it was expected, to be cared for by the LMS out of Livingstone's missionary salary of £100 a year. He then returned to the Upper Zambezi, finding on the way that his house at Kolobeng had been looted by the Boers.

Although Sebitwane had died his successor, Sekeletu, approved of Livingstone's plan to open a route to the sea, and provided him with money (in the form of ivory), a riding ox, provisions, and porters for the expedition to Luanda, Angola, and back. On Livingstone's return, Sekeletu was equally generous in giving him the means to travel east along the Zambezi to the Moçambique coast, Livingstone having made a committment about the mission to the Kololo kingdom.

When Livingstone reached Tete on the lower Zambezi, he left most of Sekeletu's men there under the protection of the Portuguese, promising to return soon to take them home. He then went on to Quelimane, the principal port in northern Moçambique, and thence by sea to Britain by way of Mauritius, the Red Sea and the Mediterranean.

During his three years of trans-continental travel, Livingstone had kept London informed of his progress, and sent the fullest possible reports of his geographical work, which included for example, the first plotting of the course of the upper Zambezi. But the last stage of his journey home was not happy — at Quelimane he received a letter from the LMS which called his whole enterprise into question; in Mauritius, Sekwebu, one of the Kololo, committed suicide after misunderstandings with British seamen; and while passing through Egypt, Livingstone learned that his father Neil had died.

1. TO T.L. PRENTICE

Barque George at sea
off Rio de Janeiro

27 January 1841

My Dear Friend,

You have returned from the celebration of Christmas and settled down quietly to your studies for some time now. But have you remembered your promise to write your humble servant when you had done so? If you have not you must understand that I fully expect a letter from you when I reach Algoa Bay advising me of the state of our dear friend C's health (I make known my expectation in order that when you receive this some remorse may seize you as a punishment for your negligence) I shall be sorry if I don't. I wish I could hear now for I have often thought of you both and wished you were here, I should have been delighted to have assisted you when passing through that grievous ordeal sea sickness. It is indeed a dolorous predicament to be in, a man and his wife and the stomachs of both making efforts to quit their bodies every time the head is elevated, is really a melancholy spectacle. I pitied but could not cure — The only cruelty I was guilty of (and I am not quite sure but I should have done the same for you, if you like my friend here, had rashly quoted the text to me before the sickness began, ('Two are better than one') was quoting the same text to him when both he and his spouse were turning their stomachs inside out into one basin — I think this may give you some idea of what you may expect. It is as bad as you can picture it in your imagination. I did not feel it. This is what I saw in others and it lasted for 10 days. I promised to tell you all the disagreeables. This is No. 1 and a regular bad one it is. I may just say if I quote any names there I hope to be kept secret — excuse my saying this. I don't think you are capable of making a bad use of any such information, but after saying this I shall write with more confidence.

Perhaps you may not be in such a pitiable state from seasickness as my poor friends Mr and Mrs Ross were, for our little vessel went reeling and staggering over the waves as if she had been drunk, our trunks perpetually breaking from their lashings, were tossed from one side of the cabin to the other, everything both pleasant and unpleasant huddled together in glorious confusion. You have been on board a steamer; that is nothing to a little sailing vessel in a stormy sea such as we had when about the Bay of Biscay, she writhed and twisted about terribly. Imagine if you can a ship in a fit of epilepsy. My nervous system not being over sensitive enabled me calmly to contemplate the whole scene and certainly I never beheld such a mess before, it might be called 'the world upside down'. The storm I won't attempt to describe. It went beyond description and I hope you will see one yourself.

But oh what power must he have who can control the winds and waves! And this is He whom we can call our Father and Friend. Precious privilege. May we prize it more. What delightful weather we have had since reaching the latitude of Madeira — 10 days after leaving Gravesend we bade farewell to the cold and now while you can't keep your fingers straight and Mr Fison has to rub his chilblains I enjoy a climate more genial by far than even summer in England. It will, however, be hotter on land. The atmosphere has been really delightful. If I were writing to C I should tell her of the beautiful appearance of the sea and sky in the evenings, etc. but you I am afraid won't thank me for such observations, therefore, I shall proceed to mention some things which I think worthy of your attention in prospect of undertaking a voyage — Get a swinging cot to sleep in instead of a mattress — a swinging tray — Take no edibles except perhaps a very few oranges or apples, some seidleitz powders and a few lemonade and soda Do. Not more than 2 trunks for the cabin. Let these contain all you require for the voyage, part of them of easy access. Have all your other goods in air tight boxes and don't open them while at sea. Some Wesleyans last voyage of this ship examined their dresses after being 3 weeks at sea, found them all right, but the act of opening them admitted the sea air and when they looked about 3 weeks after that found all of them completely spoiled. Expect to be sick and take nothing from your trunks until that is over except your Bibles; if you escape sickness it will be an agreeable disappointment.

You will require a lantern for your berth. A folio is perhaps better than a writing desk. Get your bedding and everything else arranged previous to sailing and you need not be downcast though I have told you of the disagreeable ordeal through which you have to pass. Take everything as easily as you can is the best advice I can give you in prospect of changing land for sea life —

I wrote you from the Downs. Did you get that? We had our foremast split by the wind a few days ago and in consequence have been obliged to put into Rio de Janeiro for a new one and a supply of water — We are, while I am writing, in sight of land and expect to get on shore tomorrow. We had only 30 days of water on board when the accident happened and we were in Lat.23S Long.28W. Had we pursued our course to the Cape the probability is that the first gale having taken the mast right away, should have so disabled the vessel as to prevent her gaining land for perhaps double that period. As a prudential measure the Captain puts in here for a new mast and a supply of water.

We have a good supply of tracts which we mean to distribute amongst the British shipping in the harbour and if there is an hospital for seamen I hope to find some poor fellows glad to listen to the message of mercy — if so [letter cut] please present my kind regards to Miss R when you write and tell her I have found sailing very pleasant and hope she will find it as much so. I take it as certain that she is better. I shall write Mr Fison from Algoa Bay. Remember me also if you please to your Father I would say mother too but I don't know her and Manning. I hope they all enjoy the presence and favour of God. Have the kindness to remember me in your prayers I won't forget you. [letter and signature cut out]

2. TO CATHERINE RIDLEY

On board the barque George
Lat.33S. Long 12W.

24th Feby 1841

My Dear Friend,

You must not be surprised at the liberty I take in addressing you so for I can claim relationship to you through a third party whom we both consider a dear friend. I hope you are now completely recovered from the indisposition you laboured under when I left England and I confidently expect the first letter I receive will apprise me that such is the case. But should it be otherwise we know that the hand which afflicts is guided by a heart of infinite love and that this is one of the all things which are to work together for our good. We have only to throw our whole being into His hands and all will be well.

I know you regard with feelings of interest the course which I am now pursuing as if in accordance with Our Father's will, you hope to follow in a similar track. I shall therefore tell you a little about Rio de Janeiro, to which harbour we paid a most delightful visit. Had you been there I am sure you would have enjoyed it as much as I did, even although you are more attached to home than I am. I don't think that the recollection of what you had left behind would have had even so much effect on you as it had on me for you know Scotchmen have a very great love to their cold barren mountains and warm hearted relatives. I was so delighted with the beautiful scenery, the lofty green hills covered with wood to their summits and everything else so new and strange and interesting that my thoughts never soared beyond Rio, except perhaps when I betook myself to my couch at night — The country is delightful but perhaps it appeared more so to me on account of having been for 7 or 8 weeks cooped up on shipboard. I could not cease gazing at and admiring the beautiful mountains which meet the eye in every direction and then there was so much in the people who called for observation. They are said to be revengeful and have little regard for human life. I believe it is true but I know they are capable of shining hospitality, at least some of them are as I myself experienced. As I wished to see a little of the interior of the country and nobody else was inclined to endure the heat and fatigue of a walk beneath an almost vertical sun I set off alone and having walked about 6 or 7 miles got into a Brazilian Forest. Really it is a fine sight. I have read good descriptions of them but the sight is much better. Orange and cocoa nut trees were the only trees I could claim acquaintance with. Butterflies and grasshoppers of gigantic size in great abundance. Lizards and scorpions with vegetation on a most luxuriant scale. Parasitical plants in immense variety. After spending some time admiring the strange productions of nature I travelled on and saw many vallies lovely beyond description. Their sides covered with plantations of coffee, sugar cane, Indian corn, etc. and the little shed-like cottages of the natives scattered here and there and peeping out from beneath orange trees or the spreading leaves of the banana — Being desirous to obtain some fruit I descended one of these vallies intending to seek a supply at the first cottage I came to, but no sooner had I emerged from the wood to an open space in front of one, than I was surrounded by three half starved looking dogs who seemed inclined to make an end of me but having a good stick in my hand I soon convinced them that I was not a member of the Peace Society. The row brought out the inmates and the dogs were called off. By means of a few words of Latin for I had no Portuguese I made them understand I wished to purchase some fruit, but they had none ripe. They however beckoned me to come into the cottage and partake of their dinner. The good lady pressed me to every dish they had. Beef, fish, rice, bananas, bread and cheese and tapioca, and as an inducement to eat well her husband brought out a stone bottle and filled up a glass of some sort of liquor. On my declining this he produced another and presented a glass of that. I could not tell him I was a teetotaller but kept asking 'aqua'. I never had such a strong temptation to break my resolution to abstain from these drugs as at this exhibition of disinterested kindness. Notwithstanding my partaking freely with them they would take nothing by way of remuneration although I offered to pay them in their own coin. I got the children to take a few coppers and the parents an English halfpenny as a memorial of my visit. The good lady asked whether I were 'Englees' or 'Americano'. On my answering I was the former she gave her children a long account of the English, occasionally turning round to me for confirmation to what she was saying but presently recollecting herself she explained to me that she had forgot and then again remembering that not even that was understood smiled and shook her head and when I was coming away she folded two bananas in a piece of paper and put them in my pocket. Perhaps she herself had a son in a distant land and I wished I had had some Portuguese tracts with me. May Heaven direct the steps of some missionary with the bread of life to your pretty cottage, Ye kind hearted Brazilians. The husband seemed surprised that an Englishman should refuse ardent spirits for Eng. and Am. seamen continually disgrace themselves in the streets of Rio by getting into a state of intoxication. The sailors of all other countries behave themselves like men, except ours and the Americans. They get into scuffles with the inhabitants and they soon use their knives, rob and stab them. I saw a case of this kind in the Misericordia Hospital. He was lying in a fit of raging delirium, secured by a strait jacket and the blood still flowing from his wound. The Portuguese menials understood not his language and only laughed at his vain attempts to get up. My heart warmed to my countryman. I sat on the edge of his bed and vainly endeavoured to lead his mind into a train of thought, if I got 2 collected sentences the 3rd was sure to be about blue devils or gin, etc. He was a young man of about 22 and had got this from what he called skylarking the day before. I turned away with a heavy heart for I knew from the nature and position of his wound that he could not with his system impregnated with alcohol survive another night. O How much need have the Christians of Britain to exert themselves in behalf of seamen.

On the bed from which he expected never to arise lay another who could speak English. He was an old Frenchman and had acquired the English language during the time he was a prisoner in England. He thought if he were sorry for his past sins and did as well as he could God would forgive him. I explained his mistake and gave him some tracts and left him never to see him more until we appear before that Bar where it shall be made apparent whether we have cast all our dependence on Christ or have trusted to a refuge of lies.

That you and all whom you love may meet there with joy is my sincere wish and for this end I pray that Christ may be [made] of God unto us wisdom and righteousness, sanctification and redemption.

Please write to me and ask any questions you choose. I don't know very well what kind of information you would like. Name the different subjects you should like to know and I will give you all the information in my power. Is it the difference in the mode of life? or clothing or work you feel most interested in? You know I am not very well acquainted with the feelings of those who have been ladies all their lives. A hard bed might be a greater sacrifice to one than sleeping on the ground to another.

If you ask me freely I shall tell you as plainly as I can the real state of affairs on missionary stations and what you have to expect and do. I can't of course tell you the feelings with which you will meet them. All I can do is to tell you how I like them. I gave Mr Prentice some hints about the voyage. Now I beg leave to cancel these in case you find it quite different and think I was mistaken — the only advice I shall give about it is in Scotch 'Aye put a stout heart to a steve brae'. Always put a stout heart to a steep declivity and don't be put out by trifles. I have your 'Bridges' always by me in my berth and often get my heart warmed by his heavenly minded reflections. Many thanks for it.

May every blessing attend you is the prayer of

Yours affectionately

David Livingston

---

3. TO T.L. PRENTICE

Barque George
Lat 35S. Long 3E
5th March 1841

My Dear Friend,
One thing you must not leave England without or you will be sorry for it. I mean a shower bath; you can't do without something of the sort in a warm climate. The apparatus for forming the shower will do, the rest you can construct for yourself. That with flesh brush and perhaps a pair of horsehair gloves will do more towards

the preservation of your health than all the drugs in my medicine chest. I was fully
convinced of this before we put into Rio for we have not convenience for sponging
etc. on board. Many times I have longed to have a dash into the sea but the
occasional appearance of enormous sharks prevented any of us from enjoying that
luxury. When I got out into the country and into the woods I found a fine mountain
stream came dashing down the side of the lofty hills with which Rio is surrounded.
I went up the bed of it until I got to a little waterfall. You can form no idea of the
pleasure connected [with] a thorough ablution under a natural shower bath after 7
weeks almost total abstinence from anything of the kind and many of those weeks
in a warm atmosphere — I told the Captain and Dr Grant of my good fortune on
the day we sailed. So much did they envy me that I daresay the Captain could easily
have been persuaded to remain another day in order to get one. But I unconsciously
ran the risk of getting all my clothes stolen while in the state in which Adam was
when he pruned the trees which no doubt would have caused me to look
unutterably foolish. In these forests live great number of runaway slaves, who
frequently commit depredations. I did not know this and when I did not return till
late to Mr Spaulding's all were in great consternation lest I had been robbed. A
man and his wife having gone to the same place a few days before; he was stript of
every thing he had not even excepting his wife. (Let us hope she was not an over
good one) I promised to tell you everything necessary. My mind is much impressed
with the conviction that this is absolutely so. This is the reason I have said so much
about it.

You see I have written Cathrine and sent it unsealed not because I thought you
would be offended with the liberty I take, but I don't wish to increase the weight of
this packet. I could not send it to her address, for that I don't know. Please forward
it. I did not salute any of her family because it would appear impudent in one who
had only seen them a few times. You introduced me to one of her sisters — a short
nice little creature on the evening of my ordination, but I quite forget her name.
Was it Susannah? No that is Cathrine['s] eldest sister I believe. I like her father's
fine open good natured countenence. I hope C has quite recovered from her
indisposition. Indeed I feel almost confident she has. I have thought that it would
be well for her if she has recovered to look a little into the Sitchuana Testament. I
think it would be of great advantage to her afterwards for there are many gutterals
and aspirates in the language which though no difficulty to a scotchman are a
plague to those who have not been accustomed to them. G is always gutteral, h is
pronounced as the spiritus asper of the Greeks, and n as the ringing ing a nasal
sound — 'nonarega' (to murmur) is one of the most difficult words and is
pronounced as if it spelt ngongarexa the x pronounced scotch fashion. g is always
pronounced so or like the Hebrew Heth. The other letters are sounded as in Dutch,
a as in father, e as in clemency, ē as ai in Mail, i as ee in seek, c as in church as cake
is sounded chaka, o as in pot or pole, u as oo in cool. Every letter is sounded and she
might very soon read the testament by comparing it with the English. I mention
this because I should like her to get the language well and not merely be able to say
such sentences in it as a cockney in our language 'Bring me the Hegg of the 'en' or
'You 'eet me 'ard' for you hit me hard. I should like her to practise the gutterals and

aspirates although she learnt none of the language. You can get Testaments at the Bible Society House. Please pay Mr Moffat a visit. He will give you better instructions than the above if you should consider it desirable to act on my hint. The gospel acc/ John is very easy. I have begun to make a Dictionary for my own use and have got nearly all the words in John and Acts and part of Luke.

I was much impressed with the need that exists for efforts on behalf of seamen. When in Rio they are always a disgrace to our country and to the cause of Christianity. You never find anyone drunk but an Englishman or American. Frequently they are stabbed or stript stark naked and sent so back to their ships. Yet these men consider themselves Christians. I went to two public houses where a great many seamen of both countries usually lodge and when giving some tracts to one to take on board to his shipmates whom he said were all drunkards. The American seaman who stood by said he was thankful that he was able to read tracts and requested one or two but added [to remember] 'we are not all drunkards' — I said it mattered not what we were if not Christians. He got into a rage and asked if I considered him a robber or a thief or a murderer. Am I not as good a Christian as you are? Were we not all born in a Christian country? I replied, that did not constitute him a Christian & was about to explain what it was to be a Christian but he interrupted me by saying in an emphatic tone 'I beg your pardon' — a common way of giving a point blank contradiction. As I did not wish to argue with him and about 20 ruffians around some of them drunk and swearing I gave him some tracts explanatory of what true Christianity is and told him these would inform him better. He thanked me but added with respect to what it is to be a Christian 'I beg your pardon'. I confess I was afraid of my own countrymen although they treated me with the greatest respect and thankfully received tracts and those of them who were sober listened attentively to what I said — My dear Prentice, should you be [?] going to Africa think of the claims of seamen before you settle down into any sphere of usefulness. Let that field have a little consideration when you look around for the sphere in which you can most glorify God.

We were most hospitably received by the Revd. Justin Spaulding an American Miss. at Rio. He sent an invitation as soon as he knew that there were Missionaries on board requesting us to make his house our home during the period of our visit. He can distribute as many tracts and bibles in Portuguese as he chooses without molestation from the priests but what can he do single handed among a population of from 150 to 200,000 which the Town contains, and he has to devote a great portion of his time to Eng. and Am. seamen. He received us with great affection although of course we were entire strangers and during the whole week of our stay did everything in his power to render our sojourn agreeable. It was truly refreshing to meet a brother disciple of the Lord Jesus in a place where we expected to find nothing but the heathenism of Popery. He introduced us to some other christian friends who treated us kindly and for the purpose of gratifying us they planned an excursion round a part of the Bay to the Emperor's gardens. The gardens are very fine. All sorts of fruit were in great abundance, mangoes, bananas, caju apples, custard apples, oranges, etc. The best of it however was that every one is allowed to eat as much of the fruit as he chooses. Did I neglect the opportunity Mr Fison?

I forgot to mention before I left England that if Manning is to have any medical knowledge he could have it more conveniently in Glasgow than in London. The expense would be much less and if Dr Andrew Buchanan St Vincent St. were written to I have no doubt but he would furnish the requisite information. If the whole case were stated he would be able to form a correct judgement. He was always very kind to me as the professor of the Institutes of Medicine and is therefore qualified to let you know the best course to be pursued by Manning.

We expect to land at the Cape in a few days. I write this before coming to land lest new scenes should for a little drive old friends out of my head. You can't imagine with what a light heart I visit these foreign shores. Everything is so different from the idea I formed of them while reading. The actual sight and imagination are two very different things. This is really a fine world we live in after all. Were it not that hateful rebellion against God, it would be quite a Paradise. I see more clearly now than ever the necessity of casting our whole being upon Him who careth for us. Why should we burden our hearts with these and go moping and melancholy beneath a cloud of our own vapours when he offers to look out for us. Who shall harm you if ye etc. and all things shall work together for good etc. What a capital thing that is 'all things' etc. Let Jesus be our mission's righteousness sanctification and redemption and we are complete. We shall consider it the glory of our lives to serve him in every possible way and for more love to such a Saviour as this much we have —

If possible I shall write you from Algoa Bay. I hope you have written me, advising me how you both are. I don't know why I feel so much interest in your good lady, perhaps selfishness, the prospect of having both as fellow labourers.

Affectionately Yours D Livingston.

March 17th. Cape Town — We have been 14 weeks on our passage. Fine Town this — warm but not roasting: am in Dr Philip's house. D. L.

---

4. TO T. L. PRENTICE

Kuruman
3rd August 1841

My Dear Prentice,
Through the gracious care of our Heavenly Father I have at last reached this which is to be my resting or rather halting place for a while. We left Algoa Bay on the 20th of May and reached this on the 31st July, a pretty long period you may think but through the whole it has been so pleasant I never got tired of it. My waggons are very comfortable affairs indeed, little houses in fact so you may expect to set up

housekeeping long before you are surrounded by mud walls. I could tell you a great deal about the country but any book of travels will supply a better acct. of it than I can. I go on to the operations carried on in it and first of all I want to set you right with respect to Dr Philip concerning whom I was entirely wrong when in England. I came to the Cape full of prejudice against him but after living a month in his house and carefully scrutinizing his character that prejudice was entirely dissolved and affection and the greatest respect took its place. I have heard a great deal said against him but now I am fully satisfied it is all, or at least the greatest portion of it, sheer downright calumny. He may have done some things which appeared tyrannical but it must have been from a conviction that the part he took was the path of duty and the only way whereby he could advance the interests of the Saviour's cause. Do keep your judgement suspended until you see him and if you ever have that pleasure you will agree with me that both he and Mrs P are eminently devoted and humble christians. Their worth will be known better when they have gone to reap the reward of their labours and when the name and memory of their calumniators shall be sunk in the shade of cold oblivion. He stated to me that he is only the money agent of the Society and does not wish to interfere with the modes of operation or plans of any man, but the Society had compelled him frequently to act a part he had no inclination to, by referring disputes to his decision. Whichever way he decided generally one party has become his enemy — he appears most desirous to get the cause forward and if men will only work they mey be sure of the co-operation and friendly regard of the Dr. He has been the means of saving from the most abject and cruel slavery all the Hottentots and not only them but all the Aborigines beyond the Colony. The Boers hate him cordially. Many would think it doing God service to shoot him. They have an inveterate hatred of the coloured population and to him as their friend and advocate; you can't understand it, it is like caste in India. Can you believe it? Some of the missionaries have imbibed a portion of it. I name none but you will find none of that feeling amongst the friends of Dr Philip. It is entirely confined to the [Antys] party. However you will understand what is meant by 'Colonial feelings' when you come.

I am now heartily sorry I ever retailed anything said to me by missionaries whilst I was in England. I am no partizan but I am and always have been on the side of civil and religious liberty.

I find this a pretty spot but as it is winter it wants most of its charms. The church is the largest place I have seen on any Mission Station. The walls are built of stone and so strong they resemble those of a battery. The gardens are excellent but the number of the people is much smaller than I expected to find. This being the hunting time of year most of them are out pursuing their occupation. This is a point between light and darkness. About two months hence I am going into the regions beyond in company with Mr Edwards of the station. The Church Miss. Soc. have abandoned their mission to the North of us and recalled Mr Owen one of the best of men, whose praises we heard sounded through all the colony. From all accounts he is an admirable man, every one regrets his departure. I shall see the state of the country and learn its language at the same time. It is an easy language, the only difficulty is I have to make my own vocabulary. Don't forget a good gun and learn to shoot it. Also some carpentry with tools. How is Cathrine? You did not write me at Christmas. How can I write you long letters when you serve me so? I shall be away perhaps half a year. I write this in great haste for the only opportunity of sending for a long time now presents itself [with] the return of our people to Bethelsdorp. Manning must understand irrigation. Nothing can be done without it in this country. But I can't yet speak more to him. All I can as yet say is that he might be of immense use on many stations. I shall write him when I know more accurately the [torn]

You need not expect letters for some time after this, but you must not forget to write me.

We visited Griquatown and were much gratified by a little intercourse with the brethren there. They are excellent men and have been very successful in their labours. Waterboer is a remarkable character and if you come through Graaf Reinet you will see the prettiest town in all Africa and two christians worth going a hundred miles to make the acquaintance of, I mean Mr and Mrs Murray. He is a Scotchman and is the Dutch minister of the town.

My love to Jos. Fison, up to the eyes in latinity yet, is he? This climate is most salubrious. Through the whole colony this is famed as a remedy for consumption. Several cases even of that desperate malady are said to have been cured by a residence here. So if you become ill here you must think of dying outright and not of running away to England.

This climate is certainly much more agreeable to the feelings than the English one. Remember me to your Father, Cathrine, etc. etc. Remember me at the throne of Grace that I may be kept faithful unto death.

Yours affectionately

D. Livingston

5. TO T.L. PRENTICE

Kuruman

2 December 1841

My Dear Friend,

In company with Mr Edwards of this mission I have very recently had the pleasure of visiting a large tract of the heathen country and although I have not as yet heard a syllable of you I proceed immediately to write you again. And my reason I am now doing so is I know from actual inspection more of the real state of matters here than ever I did before.

I believe I told you in my last that this is considered throughout the Cape Colony a most salubrious region so far as lung afflictions are concerned. I believe this is strictly the case. Several cases of consumption are said to have recovered by coming up from the sea coast hither. The climate is dry and to me pleasant. A great deal more so than England. But I should be unfaithful to you if I did not let you know that the summer months are very generally distressing, painfully so to females. The North wind coming over a large tract of heated country blows frequently and then the lassitude, headaches and sore eyes are as some people would say "all the go". Remember this [and] calculate upon it. Now I feel I have discharged my duty on that point. But another thing I should be sorry if you did not know previous to leaving England [?] you have heard of the great success said to have attended the missionary efforts in this locality, it is great but to one coming from England it is invisible or nearly so. Those who had to deal with Bechuanas 20 years ago and those who have seen them far in the interior alone can appreciate the greatness of the change. Don't expect to find such amiable Christians as you read of in the South Seas etc. etc. It is quite a different mind here. Don't expect to find chiefs friendly to receiving missionaries. In general they are hostile and when friendly it is generally for the purpose of "milking" them, and nothing else can be expected. It would be an anomaly if the human heart loved here what it hates everywhere else. Within 100 miles of this to the North and North East the chiefs are all hostile. They indeed receive us with a shew of apparent friendship [*del*: with an apparent shew of friendship] but they likewise tell us they do not want our gospel for it teaches men to put away their wives and this they are determined not to do and most of their people taking courage from the countenance of their chiefs are bitter scorners and opponents.

Several of these chiefs have removed farther away for the express purpose of being beyond the reach of the gospel. If you expect anything else here amongst either chiefs or people other than what your knowledge of the human heart would lead you to expect from people at home you will be disappointed. You must work uphill and work hard too and get no thanks for it into the bargain. If you come not out leaning firmly on your principles, *your principles remember*, you will both sink. There are encouragements too but these I need not mention. I shall only say that in view of both the favourable and unfavourable it is pleasant work and if I had

16

only more of the Heavenly motives within no life can be compared to it. If you come I have not the least doubt but both of you will be happy and you will continue to bless God through life that you took the step. But you will meet with things you did not expect. I earnestly hope that the views you seemed to be imbibing when I left you have not turned away your minds from your previous purposes and that the Lord will make plain your path before you. I daresay you will smile at one part of the above sentence but I feel anxious on that point and have felt some all along.

I mentioned that the tribes near us to the North and N E are all hostile to Christianity. The different villages in the other direction are not so but these are all taken possession of by the Griqua Town missionaries. In fact they have nearly encompassed us. Don't imagine I am sorry at it. I love to see their arms encircling and ready to grasp us for it will force us towards the dark interior. If no other motive impels us this will. We shall have no one to itinerate to but people of the north, there are very few people on this station and very few in the surrounding localities. Would you consider 700 as a large population for [?] and within 14 miles of this 1300 or 1400. The people are far off. There are many 80 or 100 miles to the North East, but all or nearly all are bitter opponents to the gospel.

Now I shall say something on an unpleasant subject as I believe it will be useful to you by placing you on your guard against fair appearances but you must not think I speak disrespectfully or consoriously of anyone. A most important lesson may be learnt from it. Quarrels somehow or other arose between the Griqua [Missionaries] Messrs Wright and Hughes and the Kuruman [Missionaries] Messrs Moffat and Edwards and at different times there have been squabbles and bickerings. They are all however excellent men. Perhaps the one you have seen in England is not the most favourable specimen, I have great regard for them all and their wives too are excellent characters. Messrs W and H began to employ native teachers. Whether the variance that existed had an effect in inducing Messrs M and E to adopt an opposite course and employ none I don't know but I expect that was the case. The Griqua native teachers were deficient in many respects as is to be expected from all in like circumstances; these were the subject of remark — I don't know the natives but the natives of this mission and the Griqua mission took sides. Great enmity was the consequence. The native teachers were despised by both [Missionaries] and people here and most unfortunately the heathen natives beyond us now despise them too. God however did not despise them notwithstanding all their defects he has blessed them wonderfully. Many many have been saved by their means and so great is the number of people under their influence the [Missionaries] at G can now do little else but itinerate and supervise them. When coming here from the colony we unexpectedly on a Sunday came to a village perhaps 80 or 100 miles from Griqua Town in which although from the illness of Mr Hughes no missionary had visited there for 10 months previously, school for the children and public worship for the adults were regularly observed. Many deficiencies no doubt existed but what missionary would not wish to possess a number of such auxiliaries, particularly in a country like this which has such an extent of surface. Had the tribes immediately to the North and N E of us been supplied with native teachers who could only have taught the children to read [and] perhaps could have preached in their own simple manner we should have now had a hold on them which it will be difficult to attain, for many of them are from having lived [?] become gospel hardened and stir up the more ignorant to opposition. And had the natives here been taught to exert themselves for their fellow men, I am persuaded that piety would have been deeper, they would not present such a lifeless spectacle as at present. When Mr M went to England he found the current strong in favour of native agency and he wrote to [Missionary] brethren here to employ 2. Their efforts have been blest and now the chief accession of members is from their locations etc. and those under them are by far the most warm hearted Christians. I don't believe Mr M was conscious of being an opponent to native agency. He wished higher qualifications than were to be found and the unfortunate variance above mentioned have I imagine contributed towards the course of operations adopted. Now it would pain me in the extreme if you should think I speak disparagingly of my brethren who have been in the field before me. They are all my superiors both in wisdom and piety. But looking at the subject from a different point I think it enables me to see it in a clearer light. And my reason for bringing the subject before you is I am [del: am] about to implore your assistance in behalf of the tribes situated beyond the scorners — We travelled in a North Easterly direction about 130 or 40 miles for the purpose of visiting a position of a tribe who live in a mountainous region there, a most industrious race who were busily engaged in the manufacture of wooden bowls, spoons etc. and some things iron for making picks axes and spears. The ore they extract from the mountains. A fine field for Manning for the arts are in the most uncouth state imaginable. The Iron trade is not a whit improved in their hands I should think just in the *statu quo* that it was left by old Tubal Cain of glorious memory. A furnace is built of clay similar in shape to a haycock in England having an orifice at the top and two at each end low down, a place being excavated from the ground below to enable the "blastman" to sit on a level with it. Two leather bags open at the bottom with two straight sticks fastened along the orifice and the mouth of the bags fastened round tubes of clay which enter the orifice in the furnace, constitute what may be called the "bellows in embryo". The blastman seated between these bags grasps one by the bottom in each hand over which a small strap comes to enable him to separate the opening by means of the sticks. In lifting up the bags alternately he opens the orifice of the one ascending and shuts the orifice and presses with his hand the one descending. The pressure expels the air through the tube into the furnace — By repeated strokes a constant stream of air is sent into the furnace and they by great labour raise a heat often sufficient to burn their ore to a cinder. Indeed they burn most of their ore using no limestone it is no wonder they repeatedly fail. But then they satisfy themselves it has been bewitched. They have many superstitions to prevent their own ignorance spoiling the iron, one of which is abstinence from all connubial intercourse during the whole time of the smelting and working the iron, work which engages them several months. They must not at that time even sleep in their own houses but be in the Kraal with the cattle — and no one is allowed to look at the operations. As much secrecy being observed by them as is some of our chemical works at home. They stated that the reason why we are admitted to see the whole

18

was that we were white men and had not been with our wives for some time so they hoped no harm would befall the iron. By a little knowledge of chemistry Manning might save them a great deal of iron and labour too. I told them how they might improve but of course they could not see the utility of putting in with the ore another stone in which there is no iron. The thing must be done before their eyes before they will believe white men knowing as much of the trade as they do. After smelting they weld very well but a huge stone grasped by both hands is the only hammer and another stone is the anvil.

There were 600 of them present at that village and more are expected soon as they are just returning to their former abode having formerly been driven thence by the fear of that desperate marauder Moselikatsi. This fellow is now a great way to the north but has not learnt yet to live in peace although the Boors gave him such a terrible mauling.

This tribe live in a beautiful valley situated between two ranges of mountains, the many trees in it looked very beautiful indeed after the long ride we had over the wide flat wilderness. I have not seen a better spot in the interior for a new station than this. Beside the village there is a fine fountain gushing out from beneath a rock; the water more precious than gold in this country is excellent and coming out of a defile in the side of the mountain range at a considerable elevation above the valley it could be led out for irrigation at a very trifling expense, scarcely any dam would be necessary it would only require a ditch led along the side of the valley a sufficient space for the formation of the mission gardens. The supply of water is larger than that at Griqua Town and nearly as large if not equal to that at Kuruman and the valley had more than ½ mile of excellent land (in width) capable of being irrigated. We [went up] more than 12 miles and found plenty of pasturage and water for cattle which is of great importance to every station for if the people have not these they won't remain long at any station. In the valley we saw many Rhinoceroses, Buffaloes etc and zebras and not far beyond the elephant and camelopard abound.

Now why am I saying so much about these people and their abode. The reason is they have not yet learned to despise the Gospel. They received us kindly and only know that we are friendly with them. There was only one individual who had been at a mission station and who may be called a rejector of the gospel and his influence is small amongst them. I don't mean to point out to you this as a location for your future labours. But I earnestly entreat if this is in your power. Help these people now. I know you are connected with churches which might, if they knew the immense importance of native agents, support one or two in addition to what they are now doing. They really in their own simple affectionate way can do as much towards removing the prejudices of the parents and securing the minds of the young against those same prejudices (Giving them a leaning to Christianity and civilization) as European missionaries can do for the first few years at least, and then if a European succeeds them he has a great deal less prejudice to encounter. The probability is he will be a much more useful man in the cause of Christ. I am quite well aware that you need no stimulus to generosity, I know it from having experienced it myself. But I suppose you know some churches which could do something. Could you not find some means to stimulate them. I should think few would be so callous as not to wish to be able to support a missionary of their own.

Now I have no hesitation in saying one or two pious devoted native agents are equal if not superior to Europeans in the beginning of the work. The native looks so much upon the gospel as just the ways and customs of white men that little progress is made, but from their fellow natives, the truth comes directly in contact with their minds very much divested of that peculiar strangeness which attaches to foreigners in every country; and they become teachers at a considerable disadvantage to themselves. £10 does not suffice to support a native in this country. It is only a help. They must hunt occasionally and attend to their gardens and cattle if they have any. But they take that sum thankfully and when they go hunting always provide another to supply their place. Now if you can provide yourself 1 or 2 or more by means of any church I will not count it any loss of time to see them comfortably located with some of these distant tribes and I shall if spared visit them as soon as possible afterwards with supplies etc., give you an account of their procedure and as soon as you come out I shall deliver them into your own care. O if I were with you in Gower St. 10 minutes I think I should impress your mind with the importance of this subject but the 2nd sheet is nearly finished and I have not said half of what I would.

If you had just witnessed the striking interest the natives presented as they happened to be near or distant to mission stations it would have astonished you — 250 miles off they are quite unsophisticated. At one town in particular a white face having never before been seen amongst them — they having lived out of the way of any traders, the women fled with precipitation, the children screamed with terror and the very dogs did not make to bark but ran as if they had seen a lion. But we were not long in gaining their confidence for we had two native Christians with us who explained the object of our coming and that we were people who loved all men. After the first day the women and children came to the waggons and heard the gospel; the chief of this tribe or rather portion of a tribe was the most sensible and kind chief I have seen amongst them. In general the chiefs of the Bechuanas are hereditary asses, born idiots or little better, good at nothing but begging, but neither this man nor his people gave us the least trouble. He gave us milk, porridge, boiled beans, an ostrich egg and a tiger's skin each and with the few things we gave in return he seemed perfectly satisfied and he came many miles with us in order to put us in the right way to the place we intended next to visit. Please tell C I had [the] honour of cutting out a tumour out of the hand of his [favourite] wife. She sat like a heroine, covered her face with her kaross and when it was finished looked up and smiled as if nothing had been done.

In this journey we travelled 700 miles but only 250 of these directly north, this latter distance is however farther in that direction than was ever travelled by any missionary; some may have gone farther to the North East than me, the Boors are now in that direction and have the best of the country. Towards the North West it is barren and without water.

Lions abound. We saw 6 within 20 yards of the waggon one day, but they turned civilly out of the way. They came near to us at night sometimes but did no harm more than waking us with a start. We sometimes felt the heat and want of water very much. It is roastingly hot on some of the plains, but some of the traders who go into the interior in search of ivory suffer much more than we did and for a far less noble object.

I have not yet decided where I shall place my own assistants. I am anxious to engage 2 for myself but each of these journeys costs about £5 or more and I don't know yet how much of my salary I shall be able to devote. Let me have your prayer that I may have wisdom to choose aright and that my feeble way be blest of God. I have not yet heard from you but I am living in hope that your letters are on their way.

The language is easy of acquisition. I now write addresses in it but of course there are many imperfections in them. It is much more difficult to understand the natives speak than what is printed, they do so jumble up their words together.

Please give my affectionate regards to T Fison and Manning. I should have written M now but the subject of Native Agency has come before my mind so strongly during my late journey I could not rest until I had written you about it. In case I have not an opportunity of writing M and your Father soon, please tell them that agriculture here is of the most simple description and cannot be much improved in consequence of the scarcity of water, but a knowledge of the diseases of cattle and men will be of great use to him.

At every village I am besieged for medicine, the poor creatures often suffer intensely for want of a little assistance. Just now I have 3 patients here, one a woman came 3½ miles and the other 2,140 and walked all the way and many other sad cases were presented [?] As soon as I can preach I shall visit them and remain a few days at each of the villages and perhaps deal effectually with some of them, and . . . *Post script in margin*: Dec. 15th I have secured an excellent young man. He had been on a journey of some months in search of a wife but his mind became so impressed with the duty of spreading the gospel that he hastened home and told us that there was no marriage in his head now, he could think of nothing but spreading the gospel.

He had engaged in teaching all the way home and said his heart was white, white with the work. His father an old heathen chief was angry with him next day and in order to induce him to remain offered him a garden.

6. TO T. L. PRENTICE

Lattakoo
9th Oct 1843

My Dear Friend Prentice,
'Hope deferred maketh the heart sick, but when the desire cometh it is a tree of life',
is the passage which sprung into my mind when I saw your letter. I cannot tell you
how much I longed for it nor how often I have thought of you and C. Then I
become uncharitable — then angry, then I don't know what, for I thought you
must have written and then if you had not there must be some good cause for it.
And then was that the illness of Cathrine or what? But it would be endless to tell
you all I have felt. But here you are at last. I got it on returning a few days ago from
the erection of a hut at the Bakhatla a tribe situated a little more than 200 miles
north of this. It is the nearest point in the Interior where an eligible spot can be
found, and we have by this step taken possession of it for a station. A lovelier spot
you never saw, a hill in the rear is called Mabotsa (marriage feast).

May the Lord be with the missionaries who labour there so that many may there
be admitted into the marriage supper of the Lamb. But I must return to your letter.
I need not say thank you — I am only sorry you did not send the other two you
wrote or halves or whatever they were — You ask if I don't remember old Paslow
Hall where you and I got supper one night. Remember it! Don't I! and the shake
of the hand too. But what news you give me respecting C. Has it indeed been
necessary for our Father to lay on his hand so heavily. He has seen that this
chastisement was really so or He would not have inflicted it. He doth not afflict
willingly nor grieve the children of men. O how many thoughts rise up in the mind
about you when I glance over the sketch you have given and how many changes are
now going on all around you. Salome and Miss Marshall and Radford all gone.
Lord teach us to number our days etc. But you will be more anxious to hear how the
cause of Christ is succeeding in this country than to read about events you know
better than myself. And as it may be interesting to you to know how your old friend
gets on that work I shall mix up something of ego with the intelligence.

Well then, I have been in journeyings oft since I last wrote you. These were
necessary because the Directors imagining that we should be fully employed in
acquiring the language until the arrival of Mr Moffat directed us to remain at
Kuruman for that purpose. I did not however find it so difficult as was expected
and very soon set off to gain local knowledge for myself, that being the sort Mr M
was expected to assist us by in selecting a proper site for a station, I have been
further in the interior than any other European and have repeatedly had the
privilege of preaching beyond every other man's line of things. Many merciful
escapes I have met with. Poor Birt I hear lost his wife by the overturning of his
waggon. That accident happened to me twice during one journey and not a hair of
my head harmed in consequence. On one occasion I was sitting reading with my

legs across the waggon and all I felt was an instantaneous transference from the sitting to the standing posture. I only looked back to see books, bed, bags, boxes and guns making all haste to the leeward in glorious confusion. Had my head been where my feet were a box might have finished my chapter.

Many other escapes in the field I have to thank the Lord for. Indeed they have been so many in a short space of time I now feel that I am on the everlasting arms [of my] shepherd as a little child is on the arms of its mother. No new mission has been formed in this country since I came. But we now expect quite an impulse by the arrival of the new missionaries and Mr Moffat. All is ready for the purpose at the Bakhatla and in the course [of a month or two] I expect two missionaries will be on their way to it. There are more people there than at any other spot in the interior: 8 villages on the station and perhaps a dozen round about [and this itinerary] Ross does not wish to go on in that direction so it is probable Inglis will be my companion.

We have not so many people as in some countries but this I am inclined now to look upon as an advantage, for we can bring the truth to bear on one individual again and again while in India for instance they seldom see the same person twice at services — This climate is delightful, indeed I cannot speak too highly of it. I really pity you in your cold damp one, only think I slept three weeks and 4 days at one season on the ground and although riding all that time on a pack ox I never had a cough.

But I should mention a native who travelled with me caught fever and died soon after we returned to the country of the Bakwains, (whence he accompanied me as a guide after the people of my waggon refused to go on further). I ought to be thankful for one of my party has invariably been attacked by fever [a bad sort] every time I have gone in. These don't prevail here but far in the interior. They are very fatal indeed there seems to be a cloud over the centre of Africa and that composed of pestilential malaria.

How is old Mr Ridley and Susan and all. I like much just to hear how it is with everyone I ever knew. Is Elizabeth pious now? Will you present my Christian salutations to Mr W. Ridley. Could he spend a thought or two on being a missionary to the Brazils and the immense population there?

May His presence be with your dear Cathrine and comfort her as He only can. I hope your little boy will be a missionary. Please give him a press to your heart for me. Please present my affectionate salutations to Cathrine. I can't think of her as Mrs P and will always name her as she was when I first saw her. To your Father, Mother and Brothers. Surely Fison might as easily give me a note as write some nonsense from Greek plays.

Ever affectionately yours
David Livingston

Margins:
His presence supplies the loss of all we have left behind, Blessed Redeemer help us to follow thee fully, for a kinder or better leader there never was. May his presence

[?] their false worship, if Christians could see these scenes fewer would wish to be at home. I have no doubt but they are happy in their spheres at home but they don't know the joys of obeying the command Go ye into all the world and preach the Gospel to every creature, for if they did I believe the things of mercy would soon be heard in every land. It is a pity the feeling has got abroad among the churches that to go to the Heathen includes great sacrifices. In one sense it does but in another it is not. The promise of Christ is still true, Lo I am with you, and there is no man that hath left etc etc. I shall answer every letter of yours that reaches me.

7. TO W.C. OSWELL

Kuruman
22nd March 1847

My Dear Oswel,

Having come out here on a visit and finding a direct opportunity to Colesberg, I hasten to give you a word of salutation — We heard from Captn Vardon that you had actually sailed for India so we must now think of you as again in that sultry clime, it will afford us all much pleasure to receive a note from you.

A short time after you left Mrs L and I went a long way to the Eastward. We were at least twelve days due East of Chonuane when we reached our farthest point in that direction, there we were astonished to find the Limpopo had come round and was just three days beyond us when we looked to the sun rising — When at the Bamapela we asked where the Basileka lived and the natives pointed N.W. to them and said they lived three days (native travelling) from them. The Bamapela are a little to the South of East from Chonuane, I have felt exceedingly anxious to inform you of this lest you should give your name to any map maker with the mistake you seem to have made and, therefore, begin to the subject at once — there are a great many magnetic hills a little to the North East of our station and these extend a long way Eastward — they are composed of black oxide iron and are so powerful pieces stick to the waggon wheels in travelling —

I have thought your compasses must have been affected by this cause and led you astray, if the Basileka are not more than sixty or seventy miles N.W. of the Bamapela you have not made more Northing than that from the latitude of Chonuane 24 degrees 30' — the Limpopo it is certain comes round to our Latitude for it was directly East of us at the Bamapela — it there receives a large river called the Lepinole — then another the name of which I forget and then makes a sweep away back to the North East — there is still room for discovery.

[There is a sketch map at this point in the original.]

The boors declare into the sea only a little way North of Delagoa Bay, the natives say it becomes an immense stream the "Mother of all Rivers" after receiving these two rivers.

We found quite a cluster of tribes situated in the bend made by the Limpopo — We visited three and saw the habitat of four more — one of these is that which has nearly been destroyed by the boors — the country is much more densely populated the farther East one goes than in the centre of the country — they received us like the boors with far more fear than love — the chief of the Bamapela has 48 wives and 20 children the latter in feature all very much resembling himself he is not more than 20 years of age.

[Piece torn from letter]

Cumming has shot but few elephants this year — he had a bad attack of illness beyond the Bamangwato which prevented him doing much execution. He intends to follow your "spoor" henceforth.

We found that the waggon that you so kindly allowed us to have had arrived at Kuruman before us, it had got a turn over which damaged the tent but that will soon be mended. Yokes and everything that could be stripped off were gone except the sail, this I suppose occurred after you left. We were glad to get it as it is, we needed a waggon and but for your very great kindness should have been obliged to wait and save for at least three years more. Please accept of our united and hearty thanks for the favour I hope to be able to do something towards my duty by next year — We have now resolved to move from Chonuane and if there is no better place to be found our residence will be at Kolobeng — We go on much as usual at Chonuane there all are pretty well except little Robert who is teething — Mr Ashton and I went to Sekatlong — were summoned back in a few days and though we rode incessantly came in time to look at his little boy's grave — Allow me my dear Sir to recommend the atonement of Christ as the only ground of peace and happiness in death. To this favour and friendship Mr and Mrs Moffat — Mr Hamilton, Mrs L and self unite in very kind regards to you. Should you happen to meet with Captain Steele you will oblige me by presenting my very kind remembrance — I shall write again as soon as I have any news — probably in six months from this when the bags for the season have been made up — We hear nothing about the Caffre war. Murray we hear from a son of Mr Moffat [?] London safe and sound. Do remember and write soon [?] last request.

8. TO REV. D.G. WATT

Kolobeng Bakwain Country

13 February [1848]

My Dear Watt,

My date will show you if you have not received my last that we are now in a different locality from that to which I suppose you have been accustomed. Hoping you have received my account of the removal and circumstances connected with it I proceed to mention that we still have some encouragement in our people. All our meetings are good compared to those we had at Mabotsa and some of them admit no comparison whatever. We really rejoice that we have been separated from the worthy who now occupies that station if indeed it deserves the name.

Ever since we moved we have been incessantly engaged in manual labour. We have endeavoured as far as possible to carry on systematic instruction at the same time but have felt it very hard pressure on our energies. Although I date this at Kolobeng I am now on a journey among eastern tribes one object of which is to recruit and remove the langour of body and mind which affected me before I begin the erection of a permanent dwelling. Our daily labours are in the following sort of order. We get up as soon as we can — generally with the sun in summer, then have family worship, breakfast and school and as soon as these are over we begin the manual operations needed — sawing — ploughing — smithy work and every other sort of work by turns as required.

My better half is employed all morning in culinary or other work and feeling pretty well tired by dinner time, we take about an hour's rest then, but more frequently without the respite I try to secure for myself, she goes off to the infants' school and this I am happy to say is very popular with the youngsters. She sometimes has 80 but the average may be 60. My manual labours are continued till about 5 o'clock I then go into my rooms to give lessons and talk to any one who may be disposed for it. As soon as the cows are milked we have a meeting and this is followed by a prayer meeting in Sechele's house, which brings me home about half past eight — generally tired enough, too fatigued to think of any mental exertions. I do not enumerate these duties by way of telling you how much we do but let you know a cause of sorrow I have that so little of my time is devoted to real missionary work.

I wished to come on this journey long ago but incessant occupation prevented me till December a mad sort of Scotchman[1] having wandered past us elephant shooting lost all his cattle by the bite of a small fly — a very little larger than the common house fly, he sent us an imploring letter and instead of going eastwards we were obliged to send him all our oxen and bring him back to the side of civilization. Then our watercourse burst and detained us again. We made a start last week and our friend Sechele conveyed us about 50 miles of the way. It was gratifying to see him on his pack ox hunting for us, and endeavouring to promote our comfort in various ways. We met a large party of strangers on their way to visit our tribe, this rendered

26

his presence at home necessary. He informed me at parting that he had intended to accompany me on the whole itinerary — then presented me with about 4 gallons of porridge; and two servants to act in his stead. He is almost the only individual who possesses distinct consistent views on the subject of our mission. He is bound by his wives. Has a curious idea — he would like to go to another country with the hope that probably his wives may have married others in the meantime. He would then return and be admitted to the Lord's supper and teach his people the knowledge he had acquired. He seems incapable of putting them away. He feels so attached to them and indeed we too feel much attached to most of them — they are our best scholars — our constant friends. We earnestly pray that they too may be enlightened by the spirit of God — On the subject of admitting a man who has more than one wife I have felt as you but though not clear that I ought to refuse I am glad I have not been asked to baptize a polygamist. I think I should decline because of the evil which a contrary practice to that which has been followed if now introduced would effect.

We have monthly communions. I feel strongly inclined to weekly and will probably follow out my conviction on that point. I have many things to tell you but fear I should fill my sheet with matters of but little interest [to] you if I went on with ego.

We heard with pain that you had been called on to part with your partner soon after your union. The time which elapses before a letter on the [death] of any of our connections can reach its destination makes me afraid to say anything to those in whose losses I sympathize for fear that my letter may only have the effect of opening up the wound again.

I cannot presume to give advice because I do not know the circumstances in which you may be placed. Having been married the conjugal state may again be desirable but I would not go much out of my way for that object. Commit thy way unto the Lord. Cast all your cares upon Him. In a few short years at most we shall wend our way to the regions of eternity.

I have lately been much impressed by the thought that the first Salutations of ministering Angels to the soul rushing out of its falling tenement of clay will make it feel at home and then the confirmation from the Lord Jesus himself will certainly wipe away all tears, sorrow and sighing will flee away. Let us live as seeing Him who is invisible. The world to come exercising a constant influence on our conduct. I hope you are favoured with health in your new sphere of labour. I addressed you Almorah. Hope you have not stopped short of it. A younger brother of mine in Ohio wishes to go to China. This is his last year at Oberlin. He wishes to be connected with our society. Have written the Directors on the subject.

He does not like the American Board because of its connection with slavery and polygamy in some of its mission churches among the Indians.

He is 'engaged' to a descendant of the Puritans in New England and I suspect a bit of a blue for she has been educated at College but in this I may be mistaken.

I have felt more than ever lately that the great object of all our exertions ought to be conversion. I could not reasonably expect it until now but I think we may now hope for it for knowledge is increasing among the Bakwains. But nothing will induce me to form an impure church. 50 added to the church sounds fine (at home) but if only 5 of these are genuine what will it profit in the great day? In some parts of this country where the gospel has been preached for a number of years nothing would be easier than forming a church like [torn] [for] relief or by secession but that won't do for Scotch [Independ] nor I fear will it do for the poor people themselves at [?]. Mr Moffat and Ashton are busy printing Isiah. We have finished the proverbs, preacher and when Isiah is finished will issue the three books together.

The Caffre War is not finished yet. All the missionaries have been driven helter skelter out of the country; this I don't understand — the mission premises are burned I think I should stick by my people to the last but that you will say is the same folly as to get bitten by the king of beasts. The brute I rather call the villain. I like to promote their interest and shew them that [I love] them.

Lately we were menaced by an attack by a party on horseback. I got information of it one night and instantly went up to the town and when our invaders came all were ready. The worthies thought proper to deny that they had any bad intention but having come at a time when horses die, the loss of a portion of their horses shewed they had been baulked of their intention. They would never have risked their horses had they not expected to carry off cattle. Some of our old big-wigs said to me 'Now we understand you are a real friend and would die with us'. Our missions do not seem the right sort of thing for the world we live in. When a station is formed and people converted then great expectations are formed of what [to] be done with the young — but the young grow up in the belief that a supply will always be had from Europe and though a missionary die another supplies his place with the express intention of devoting his energies to the young, those who have not felt the evils of the old system never stand forth as propagators of the new. No native has ever in this country taken charge of a station formed by a European. The converts who show that theirs is a religion for all circumstances in Tahiti are not the young in general but the old who felt the evil of the old system. I begin to believe that more attention ought to be bestowed on adults than on the young. What do you think? Our projected Institution will I fear not be established soon, the opposition I met seems to have damped others — all our missions are too far South. The real heathen are near and beyond us but what can be done? I have written the Directors on the subject but who am I, a poor thing only that can write to nobody but yourself. I have given up McLehose on account of what you said. Unless he writes me I can address him no more. It is his fault but I am away from the subject. The Directors can do very little. They appoint but we do not go to our destinations. One sent to China or India I forget which has taken unto himself a wife and sat down at Hankey. By the way as I write you in all confidence what do you think of Hankey Tunnel? A body of converted Hottentots need an improvement to their station, the missionary nobly devotes his energies to the undertaking $100 is granted 'out of a certain fund', contributions are levied among the good and perhaps another $100 is raised. Now besides the purchases of tools and gunpowder every mother's son of these established Christians expects and receives daily wages for that which is intended for his own benefit. Since the thing was for their own benefit why do they not make their own tunnel. Our people made our waterxourse and school although heathens and the former entirely for my benefit. There may be something else in the Hankey tunnel as I hear something of a college there and you may be able to inform me as I feel perplexed in some things.

Yours affectionately
D. Livingston

1. Probably Cumming q.v.

9. TO DR J.R. BENNETT

Kolobeng, Bakwain Country,

23 June 1848

My Dear Friend
Your kind favour of Sept. '47 lately reached its distant destination. It was preceded some months ago by another accompanied by a parcel and though I have been unable to acknowledge them sooner, I have not been insensible to your kindness. Both letters and parcel have proved great treats to us in our solitude and I wish mine may only afford you half the pleasure yours imparted. We have been so fully occupied since our removal to this locality in erecting temporary dwellings and then more permanent buildings — clearing land for corn and teaching I could not allow myself the pleasure of correspondence. Every day is spent in somewhat the following manner. We rise early and hold school, then manual labour as hard as we can work continues up to the time when the sun declines. I then go to the town to spend an hour or so in conversation with any one willing to be taught. We have three week evening meetings as soon as it is dark and a prayer meeting every evening in the chief's house somewhat later.

My better half has an Infant and Sewing School immediately after dinner attended by from 60 to 80 children. As the only intermission of my duties through the day is about 15 minutes for dinner I think you will excuse me for not acknowledging your favours more promptly, the manual labour will be less severe by and by. We are at present finishing our new house. It is easy to build a castle in the air but no joke to build a cottage on the ground. One must begin at the beginning of everything here and be present at the beginning, middle and end of every operation — the natives work willingly but make everything round — even graves are of that shape — set them to build a house — walls, doors, windows, fireplaces, everything would be round. The propensity to rounding in all their works is so strong in Bechuanas very few can be broken of it. The seams of their carosses are straight but there are but few cutters in a tribe — The "Niger Expedition" and other pamphlets were extremely welcome and fully answered your kind intentions

29

in sending them. Many of the topics adverted to in your last were quite new to us and awakened feelings of interest agreeably different from those which usually occupy our minds — the new planet[1] — the new view of the Pleiades — Etherization etc. If nearer to you I should be tempted to plump the question — have they seen the lost Pleiad? How strange that the centre of the system should be less visible than the planets. There is much to cheer amidst all the strife in which you live. Here afar from the din of war and strife of tongues I feel more inclined to rejoice over the wonderful goodness which God seems to be working out for this 19th century than be depressed at the bitterness and folly displayed by some of whom we might have expected better things. You ask why I do not immediately repair to the Cape for medical advice. It was impossible to travel in our jolting waggons till it was becoming better. It would have taken at least three months to reach it and more than six to return, then we were beginning an infant mission — my companion very unsuitable for such an undertaking in consequence of excessive irritability of temper. I have always had an aversion to returning even colonywards, all my desires are to the North. A wish has never yet crossed my mind to return to England. I fitted a screw to the arm so as to produce pressure between the ends of the bones and thought the ligamentous union might become osseous and by the time I became undeceived the arm becoming useful. It is now nearly as much so as the other only more unsteady. When making the roof of our house at Chonuane a lathe on which I was standing broke and I became suspended by the weak arm and swung like a pendulum but both on that occasion and another in which I fell from the beams a distance of about 9 feet the pain in the part was only temporary. I have no periodical pains as is usual in lion bites. A man who was bitten at the same time had his wounds burst open at the same time in the following year — and Mebaloe the native teacher feels a numbness in his wounds occasionally. I shoot with a heavy rifle but with the left eye and the right arm extended as well as before the accident. I lately shot an enormous buffalo at about 20 paces from our door — many shots were fired as he rushed through the town and at full speed down to us but mine went through the heart. You will perceive by this and my constant manual labour that it is in mercy still a useful member. I fear for the consequences however in old age should I be spared to reach it and were you to come our way I should gladly avail myself of your assistance, even without ether. I often saw you in my dreams when in a state of extreme helplessness and when you began to undo the bandages instead of London I was in a miserable little hut erected for me by a kind native. From the above incident you will perceive we live in a wild country, buffaloes and rhinoceros have by mistake come dashing through our town eight times since we came here but all were killed. We do not feel uneasy nor have many of the feelings which when at home we thought of these things — I had a great horror of serpents but usage not courage makes me care but little for them now. We killed one 8 ft 3 inches long by our house here a few months ago. The clear poison dropped from its two upper fangs for a long time after it was beheaded. Another about 6 ft more recently and one evening when returning from a prayer meeting I put the lantern on a box at the door and looking down saw one — a cobra moving just at my toes. Had I seen this in England I should have been petrified but men are so wisely constituted by the author of our frames, the mind as well as the body readily acclimatize and many of the natives are bitten by serpents but few deaths occur from that cause. Suction with some soft bark of a bush interposed between the tongue and wound is their chief remedy. I have introduced cupping over the part — the first bite I saw that of a puff adder I treated by incision and scarifying the limb afterwards to relieve the enormous swelling, medicine etc. but since I have seen native doctors more successful without cutting out the part. Lately, when I was on a visit to some other tribes Mrs L saw a serpent in her bedroom; as it could not be caught the thoughts of such as uncomfortable companion almost robbed her of sleep. But I need not mention these things. I suppose you have a pretty good idea of what the country is. I am sorry you have not received any specimens of the productions of this country, I shall begin and collect some as soon as we get into our new house in a week or so. I shall write you when I send them off and hope they may reach you in safety. We often lose articles on the way between here and the Cape. At present I have nothing except specimens of the geology of the country and these are not fossiliferous. Our summer begins again in about a month hence. Spiders are to be found in greater numbers and beautiful variety than any other except the grasshopper tribe. All insects must be preserved in spirit. The climate is so dry they become brittle and the jolting of the waggons breaks them if preserved as usual. I shall not be able to make up a box in less than six months but I shall answer your other letter in the interval. The reason of our removal hither was the want of water at Chonuane for irrigation.

Here we have excellent water — a great blessing indeed and should we have peace for a few years we shall be surrounded with many comforts in consequence. Nearly all the English vegetables grow well in our winter and peaches, apricots, oranges, apples, grapes, figs flourish in summer. I have seeds of all these in the ground and some are two feet high — there are some native fruit trees worth transplanting and I mean to try. We have an olive tree about a foot in height and ginger (one plant). Some of the medicinal plants might succeed if we had fresh seed to try. Rhubarb would be of great value. Also jalap — none of the medicinal seeds though vegetated. Could you procure any fresh seeds of the medicinal plants? We should feel very thankful for them. I have castor oil trees in the garden and use the seeds but they cause great nausea before they operate.

We have much cause to be thankful for the favour we possess among the people. All treat us with great respect and many shew much kindness — But there is no Southseaism among them. Those who are well disposed persevere in learning to read — they act with great caution — so conversion is a plain prosaic affair among them. The chief and principal men are our best scholars and possess more knowledge than any of the poorer people — the former reads a great deal is never absent from a single meeting and invites his people also to attend. We know a great difference in him since we came to him. He was a rain maker and had the reputation of being a wizard. But he put others to death for witch-craft and thought it meritorious in him to do so. He has nothing now to do with the rain making incantations and it was by his own desire we began prayer meetings in his house. He told me on that occasion that he always prayed in secret and wished to have

31

prayers in his house regularly in order that others of his people might be induced to believe. Polygamy is his great obstacle. He asked me to send him away to some other country in order that he might without distraction attend to instructions for three or four years and probably his wives might in the meantime have become married to others. He would then be admitted to membership. We can sympathise with him for his five wives are the most agreeable females in the town. Observing soon after our coming to live with him that we placed much importance on believing in Christ he several times offered to thrash his people into belief adding you who are kind hearted people do not understand how 'hard' hearted they are. I can do nothing with them without beating them. Now, however, he invites them to come to the services and argues with them and is much milder in his government. He always appears clean and dressed, purchases soap, makes candles and admires everything European. We entertain hopes of others in the tribe but have need of patience. The people behave with decorum in the chapel — seem to listen attentively — the numbers are from 100 to 150 on sabbath. Attendance on the infant school being in no way compulsory we hope good is being done, all who come do so simply because they like it and yet there are seldom fewer than 50 and sometimes they amount to twice that number. It is very different work at Mabotsa. Soon after we left some unpleasantness arose between the missionary and chief. The former had to pay a fine to be allowed to remain and as a heathen does not readily forgive for many months afterwards a single Mokhatla never entered either school or chapel. The benefit of the mission has been ever since almost entirely confined to some families we brought from Kuruman. It is a great pity for it is one of the finest localities in this country. Never send any artizans as assistant missionaries. My companion at Mabotsa was a good carpenter but from the existence of a feeling in this country between ordained and unordained missionaries similar to that of the clergy to dissenting ministers he was determined to show me his superiority and I had to set every stone and stick of my house square with one arm and the other in a sling. I felt ashamed to tell anyone of what I had to do because ordained and that my neighbour might not appear as my servant in the eyes of the natives. I made the watercourse and was the real servant. I never mentioned even to my (about to be) father in law. But my colleague would not rest satisfied with his standing till I had given him the station entirely. I left a good house and garden in excellent order — my friend at the time had neither built a house and school at Chonuane. My lips were either blisters or scab all the time by the sun it being the hottest season. I burned them lately to get the nails out of the roofs and now we have got up another house and prepared another garden and hope to be allowed to devote my time more to missionary labours. I mention these disagreeables which I have met because they have had a bad effect on my heart — I often think I have forgiven as I hope to be forgiven but the remembrance of slander often comes boiling up although I hate to think on it. You must remember in your prayers that more of the spirit of Christ may be imparted to me. All my plans of mental culture have been broken through by manual labour. I shall soon, however, be obliged to give my son and daughter a jog along the path to learning and as you very kindly request me to inform you of what we need I shall mention elementary books such as you judge best for your own children.

I hope you will be able to read this without difficulty my hand is unsteady after working all day. If any sentences appear non-sense from my ideas remaining in the Sitchwana idiom please transform them into sense. I shall do better or try to do so next time. I never had time to take the latitude till a few days ago and that was in consequence of having given my leg a cut with an axe. It is 24 degrees 38′ S. and we are about a degree and a half west of the longitude of Mosega which you may see on maps. I have not the means of taking the longitude.

Your family increases very fast and I fear we follow in your wake. I cannot realize the idea of your sitting with four around you and I can scarcely believe myself so far advanced in the world as to be the father of two.

Please present kind salutations to Mrs B for us and believe me.

Yours affectionately
D. Livingston

I need scarcely add how glad we shall always be to hear from you — and by Post.

1. Neptune q.v.

10. TO MISS INGRAHAM

Kolobeng
20 May 1849

My Dear Sister,
You must not be alarmed at seeing a "nigger" rushing so unceremoniously into your presence and presuming to address you by the above title. I might give some account of myself by way of introduction but as I neither like telling you how bad I am nor yet sounding my own trumpet and still wish to break the veil for myself, you must bear with my rudeness. The lock of your hair Charles has kindly sent us makes my heart glow with friendly feelings toward you. And I needed something of the sort for I have imbibed a little prejudice against some people on your side of the Atlantic. Indeed I owe them a grudge on account of the monomania under which they labour touching the complexion of my adopted country. I suspect Pocahontas never knew one afflicted with the malady [?of] her time. Is that then their way of "going ahead"? Has the grievous loss we sustained in the extraction of so much worth as went away with our Ironsides and sturdy Independents been all thrown away on their progeny? I feel quite crusty towards them when I remember their privileges and often wish they would recollect what is expected of them in the nineteenth century. But you I am sure are not one of their number. A descendant I

suppose of the Plymouth Fathers. I take it for granted you have much of their spirit and salute you as beloved for the Father's sake.

There is however one item which before going farther I wish to settle with you. It accounts for what you may imagine the oozing out of the foregoing bit of spleen. It is on my mind and I must get rid of it by submitting it to your consideration. You love your parents dearly and fear should it be the will of God that you go to labour for him in a foreign field the shock of your departure would kill the worthy couple. Now my dear sister if they are Christians the shock of your departure is the very thing that would renew their youth. I am willing to be called a false prophet as long as I live if the effect you fear follows. It never occurs in England. Why should it in America?

I know of several cases in which the refusal of the parents to allow their children to go has been followed by the death of the latter. And I shall tell you of a missionary (lately deceased) who was nearly but not quite in Charles' situation. Two sisters were married to missionaries from one family. The mother was dead some years before, so the father resolved to keep the only remaining daughter to nurse him in his old age. The above referred to missionary went on a visit and the old man suspecting the object of it tried several ways of getting rid of him, but my friend continued to ply his suit for a fortnight and was as he ought to be completely successful. The old man is still I believe alive and well and never regretted the loss of all his daughters. Nor will your parents regret dedicating their child to Jesus. But they ought to be tried. Charles has two strings in his bow when he has you on his side and I shall not think you have fairly plied them unless you have poked at them for at least a month. As for yourself, there is a sweet promise concerning whosoever shall leave Father and Mother and there is not the shadow of a shade of doubt but it is as true as it is sweetly consoling.

But as the question as to whether it is the Divine Will that you go to a heathen land remains to be decided by yourselves before Him who knows all our motives. May He be your guide who cannot err. Here we have a hard field to cultivate. The people are slow of heart to believe. They love everything but the truth. Our hearts are often sore when we think on their state and prospects. When we hear the loud wail of death in the town it distresses us for we know that there is neither ploughing nor reaping, dancing nor singing, in the region to which they are borne. They appear to love us and are complacent towards everything except Jesus and the resurrection. When will the dry bones live? Pray for us my sister. We shall never meet on earth but we may hope to see each other in Heaven.

The chief of the people is the only individual in the tribe who professes faith in Christ, is a very superior person compared to his people. When I first preached to him he asked me why it was my forefathers did not come to tell him of judgement and eternity. Why had they all been allowed to perish. And one of his people said to me afterwards "When we stand together in judgement I shall tell God that you did not teach us, you only came to us once".

These incidents seemed to indicate that I ought to come to them. I have never regretted taking up my residence here but it presses heavily on my spirits that their

feelings are so evanescent. I have hopes of the second individual above referred to.
I pray that God may indicate your path plainly.

My better half is gone to Kuruman or Lattakoo on a visit to her parents. Were
she here she would send love as hearty as my own. We have American missionaries
on the Eastern coast who have been very successful in their work. We do not come
into contact with them — they are more than 300 miles distant and the intervening
space is unsettled in consequence of the lawlessness of rebel Boers.

I should like to know them for though I feel inclined to say sharp things to some I
have a warm corner in my heart for American Christians. You must excuse my
want of politeness in this my first attempt at addressing you. I am so far out of the
world and am among complete Barbarians. But I love Charles much and shall feel
pleased if you can only gather from some parts of it that I love you for his sake.

> Believe me my sister
> Affectionately yours
> David Livingston

---

11. TO W. FAIRBROTHER

Kuruman
14th January 1851

My Dear Friend,
I had almost concluded that your bowels of compassion for us poor Hottentots had
got into a state of hopeless constipation when I was cheered by some small
symptoms of amendment in the shape of a note, you dared not call it a letter and I
am glad of it for if you had I should have set you down among the great potbellies
who modestly say "I shall give my mite" and the mite means the same thing as it
did in the case of the old man you and I saw at the missionary meeting at
Brentwood (or ford) so carefully fishing out five shillings from among the
sovereigns — But you promise amendment, very well, better late than never. I
write you now as a matter of conscience for I think I answered your note after
receiving it about two months ago. I am not quite certain but I wish to make sure
and I have very little to write about. We came out here on account of my wife's
health, she was troubled with an affection of the nerve of motion of the right side of
the face. It became quite paralysed but I am thankful to say it has regained its
power. We are proceeding to the Cape for the sake of surgical assistance. I require
my uvula excised, but we were prevented from proceeding and now we hear our
station is again menaced by the Boers and I feel anxious to be back again. They are
great plagues to the progress of missions these same boers. They are of Dutch

extraction. Each has his big bible which he never reads. Each has his horse and gun with which he can kill the blacks. They look upon themselves as the peculiar favourites of Heaven — That they resemble the children of Israel when led by Moses. And the blacks are the descendants of Cain and may be shot as so many baboons — Sechele once warned one of them to depart out of his country — this as the French say created a great sensation among them. It was worse than Balaam's ass' palaver with her mad jockey. But the worst feature in their case is they all believe themselves to be Christians and the miserable delusion is kept up by their ministers who I am not proud to say are generally Scotchmen. I saw two of them called Faure and Robertson baptizing their children by hundreds. They did not like to see me and when they returned to the colony propagated some slanders which they had collected among the boers against myself. We hear that the natives and boers have got to fighting — two have been killed on each side. I fear the natives will ultimately go to the Wall. They cannot stand against the European — they must ultimately go to the Kalahari desert and the regions where fever prevails so much that Europeans can't live. Providence seems to reserve Inter-tropical Africa for the black races. We have good hopes of finding an opening into the new region beyond the Ngami and hope to proceed thither in the course of this year — The Portuguese slave trade extends right into the centre of Africa. Sebitoane who is the great man in the new region was one of the immense swarm of Mantatees which came down and was defeated by the Griquas in 1824 but beyond him there is a tribe called by the euphonious epithet of Babolebotla and all, both men and women, knock out the upper front teeth on attaining the age of puberty.

There is a lamentable deadness prevailing in the whole of this field — There are no conversions the only work proceeding with anything like briskness is the translation of the bible into the language of the people. Mr Moffat may complete it by the end of the year and then if spared he may say he has not lived in vain. The Sitchuana is extensively spoken. A few words quoted by Dr Krapf incline me to believe that the language there spoken is a dialect of the Sichuana but I cannot well judge. If you could either beg borrow or steal a copy of his dictionary from the church missionary society I should feel much obliged to you.

I wish you every success in your present occupation — May you have grace to preserve you in the right path. Never forget the end of all things which is at hand. The Lord bless you.

Mr Hamilton is in a state of extreme debility — walks on crutches, cannot see to read. I hope I may not live so long — All the missionary party here are well. Accept kind salutations from my wife and self and believe me ever

Affectionately yours
D. Livingston

Please address in future D. L. care of Mr David Arnot, *junior*, Colesberg, South Africa. Don't address them to the Cape.

12. TO W.C. OSWELL

Saturday evening.

[15 November 1851?]

My dear Mr Oswel,
We have wandered — by taking a broad waggon path made by the Griquas — I was so much taken up by some newspapers we found at Sekhomi's that I paid no attention to our course till too far on to retrace our steps — we left this morning and had we gone across back again [sic] the country we could not have reached Koribelo this evening. So I have resolved to send a man to you tomorrow and if we can find a guide at Shokotsa or Serinan we shall cut across the country so as to reach you before you leave Boatlanama.

[There is a sketch map at this point in the original.]

The relative position of Mashue and our waggons is not correct in the above scrawl, Mashue is more to the north — but you will understand my proposed route by the above — when I first perceived about mid-day that we had taken another path than that to Mashue I saw we should be obliged to make about 8 miles of Northing before we could reach the road again. We are in the Shokosta road. The Griquas have gone by it and we saw some people coming back from — their waggons with dried meat — they report that the last Griqua waggons left about four days ago — they have been killing eiland.[1] We reached Sekhomi's on Thursday midday and would have left on Friday afternoon but our worthy Maimeloe strayed or slept with the cattle and coult not be found till near sunset. The papers sent are all I found for you. Those I got are Cape papers of Feb and one of March — the news of course just what you gave us in April. By a private letter I hear that Mr Murray has published his work with the title "Discoveries in South Africa", price 24 shillings. Hope yours will be cheaper, but lest it should not I may as well let you know that I shall expect a copy gratis. A letter to Mr Rutherfoord from yourself is published in one of the papers — about the Zonga R. of last year — and another I think from Macabe about the Limpopo which I shall cut out for our friend Ollareton. I feel quite sorry for Sir Harry Smith, never were hopes and plans more completely crumpled up by unforessen events than his are — He deserves the sympathy of all, but sympathies don't seem to flourish on colonial soil — they abuse him right bitterly — Missionaries come in for a large share of abuse too. Mr Godlonton [4 or 5 words obliterated] with rage at our squad. Most of us, however, are able to bear such ebullitions with equanimity.

The Caffres have undoubtedly the best of it. They have beaten the Colonists, I mean, and reduced them to a persuasion that they are helpless without regiments from home — but they don't like to admit so much, and hence feel disposed to blame everyone else.

The tribes south of us are in a disturbed state in consequence of the doings and sayings of the Boers under Pretorius. Sechele sends a message by Sekhomi to the effect that it will be proper to send a man forward on horseback from Logagen to ascertain from him if it may be safe for us to come out. The Boers have, it seems, been planning to place spies at, or near Kolobeng, to watch for Englishmen coming out. But most of their plans and threatenings are mahuku hela.[2] Basiamanga is here — came yesterday. Dickie arrived the day after he left. Kamati or Kobati left the gun at a cattle post; Sekhomi promises to seize it — offered to send it after me to Kolobeng, but I requested him to keep it till we came in again — I could not wait till he should send for it. He sends you salutations.

We have got two goats and a sheep — these will last some time. We did not see a single head of game except a few pallahs[3] — so we have not regretted that we declined your kind offer of the horses — I told George that my reason for declining them was the fear that if we killed anything it would only delay us and our wish was to go along as quickly as possible. Hoping to see you again at Boatlanama, I am

Affectionately yours
D. Livingston

As it is just possible we may not get a guide across the country please do not wait any time for us at Boatlanama. We shall move slowly. The man who takes this is paid. I beg a perusal of your papers when we meet again.

1. Eland.
2. Nonsense.
3. Impala.

13. TO REV. R. MOFFAT

[En route to Cape Town?]

January 1852

. . . you ought to take into consideration whether remaining at your post in your present state of health is not treating your animal spirit rather unfairly. If it were possible that you could barely live till you had finished the translation then I should be content that you push on night and day with it but continued present application to that work may shorten your period of usefulness in the world and I am sure that you are needed yet in the present dispensation of riot, rebuke, blasphemy, adultery and nonsense. No one would like better to see you persevere in the translation till completed than I, but it admits of question whether present application will not rather hinder the desired consumation than otherwise.

Present our united salutations to Robert and Helen his spouse, as the gravestones in Scotland say, and MaMary, the children send theirs too. Thanks for the manuscript. I actually looked twice through the bundle of papers in which it was found and did not notice it. I must have done it mechanically while thinking about something else. If you should write to the Cape please enclose a scrap out of the Natal Paper on the account of Evans' speech in which he or some one else refers to "English Capital".[1]

If you think of anything we can do for you at the Cape you may still let us know by the post, Mebaloe talked of the iron axles but I did not encourage him to think of them further than saying they were for sale.

Believe me affectionately yours
D. Livingston.

Zouga's eyes still plague him, a tooth coming keeps up the irritation, the rest are all well.

1. See Appendix I.

---

14. MR JOHN SNOW PATERNOSTER ROW LONDON

Litubarula
Bakwain Country

12 Jany 1853

Mr Snow.

My dear Sir

The last parcel of books came to hand in a damaged condition in consequence of having been done up in a regular dishclout fashion. The plates of Maclise's work were damaged by water and the covers of several of the other works had been soaked probably more than once. I had imagined that the natives of Paternoster row knew better than send books abroad wrapped only in a bit of brown paper without the teaching of a missionary. You do not mention whether you drew the sum £10 I authorised you to do. It you did, please put it down in your bill on the proper side and let me see exactly how matters stand. Then furnish the following order and draw, for the amount. If you show this to Dr Tidman he will allow me to draw, say £10 — but I must know the exact sum in order to advise Mr Thompson of Cape Town of it. Please attend to this, and if you don't should I be cut off by the Fever you will never get a farthing from my executors and indeed would not deserve it.

Bloomfield's supplement to his Greek New Testament.

Laycock's Analeptics — was reviwed a few years ago in Forbes' Review — if not published by Highley or Churchill, it may have been at York as Dr Laycock lives there.

Moral aspects of Medical Life including the "Askesis" of Prof Marx. C. Churchill & Stephenson.

Evening thoughts by a Physician (London 1850).

Daniel on the diseases of the coast of Guinea. Highley.

Boyle's diseases of Western Africa.

Bryson on diseases of Western Africa.

Principles of medicines and therapeutics (*Latest edition*) by Dr C. J. B. Williams

Maunsell & Evanson on children Fanuel & Co.

The following volumes of Sir W. Jardines Naturalists library.

Birds of Western Africa 2 Vols — Ruminating animals — Lions & Tigers — Elephants & Rhinoceros — Sun birds — Fly catchers — Entomology.

Every day Wonders or Facts in Physiology etc.

Uncle Tom's Cabin by Mrs Stowe.

Gesenius Hebrew Lexicon — There is a late edition with many improvements published in America — should prefer it.

The remaining two vols of Bunsen's Egypts place in universal History

First Vol of Prof Owens lectures on the invertebrata.

If you can get Hippocrates as printed by the Sydenham Society *second-hand* I shall feel obliged.

A few more fasciculi[1] of Maclises have come to hand. Send the whole.

To the above may be added a septuagint of clear type of the letters about the size of those employed in Bloomfield's Testament but the volume itself not very large.

Believe me yours truly
David Livingston.

1. *Latin*: bundles (of pages)

15. TO HON. R.W. RAWSON

Cabango

18th May 1853

My Dear Sir,

I am sorry I cannot give you the route thus far but am nearly blind by a blow on the eye by a branch on riding through forest. I have given Quango and will add others which I have rendered, reserving the full observations for the Kuruman route. We went more to the east this time, crossing Chikape R in Lat 10° 10′ and Long. 19° 42′ E then Loajima R. in Lat 9° 58′.

Had fine observations at another river called Moamba 9° 38′ S and Long 20° 13′ 30′.

From this I deduce the Longitude of Cabango which is on the Chihombo as Lat 9° 31′ [?and] 20° 31′ or 32′ the moon being now too near the sun.

All these are considerable rivers containing hippopotami and must be crossed in canoes. The Casai is about 40 miles East of Cabango. We shall cross it too in going south which we shall now do at Katema.

Had a terrible attack of Rheumatic fever from sleeping some days on a plain on which the water was flowing ankle deep. We had trenches round our berths, but I had 25 days of it and am now very weak having lost much time besides.

Excuse me this and
believe me Dear Sir,
Yours sincerely
David Livingston

PART TWO

The Zambezi Expedition
1858 • 1864

CHRONOLOGY

1858	March	Zambezi Expedition sailed from England: David Livingstone; Norman Bedingfeld (naval officer); John Kirk (physician and botanist); Richard Thornton (geologist); Thomas Baines (artist and storekeeper); George Rae (engineer); Charles Livingstone (moral agent and photographer); Mary and Oswell Livingstone
	March	Mary pregnant, left with Oswell at Cape
	April	Arrival at Zambezi delta
	July	Departure of Bedingfeld
	September	Arrival at Tete
	November	Kebra Bassa gorge
1859	January	Journey up Shire River
	February	Return to Tete
	April	Visit to Lake Shirwa
	June	Return to Tete
	September	Visit to Lake Nyasa (Malawi) Baines and Thornton dismissed
	November	To Zambezi mouth to await news of extension of Expedition
1860	April	Revisited Sekeletu, bringing some porters home; discovered death of Helmore and LMS mission
	September	Return visit to Victoria Falls
	October	At Kariba
	November	At Kebra Bassa

1861	January	To Zambezi mouth to await new ship *Pioneer*
	February	Universities' Mission to Central Africa arrived at Zambezi mouth; journey to Rovuma to find alternative route to Shire highlands
	March	Visit to Comores
	April	Return to Zambezi; UMCA mission taken to Shire highlands
	July	UMCA settled at Magomero with freed slaves; armed foray against slave raiders; Livingstone on Lake Malawi
	November	Return from Lake
1862	January	Mary Livingstone and missionary ladies at Zambezi mouth; *Lady Nyassa* delivered in sections; missionary ladies taken up Shire *en route* to Magomero; Mary left at Shupanga
	31 January	Mackenzie died, followed by Burrup
	March	Missionary ladies taken to Zambezi mouth on way home
	April	Livingstone returns to Shupanga
	27 April	Death of Mary
	June	*Lady Nyassa* launched
	September	Second trip to Rovuma
1863	January	*Pioneer* tows *Lady Nyassa* to Shire cataracts
	March	Thornton's return to Expedition
	April	Thornton died; Charles Livingstone and Kirk left Expedition
	June	Arrival of Tozer to replace Mackenzie
	July	Expedition recalled with effect end of 1863; *Lady Nyassa* not launched on Lake, but Livingstone sailed along west coast, then on foot to watershed past Kasungu
	September	Return to Shire
	December	Tozer withdrew UMCA to Zanzibar
1864	January	Livingstone returns to coast with *Pioneer* and *Lady Nyassa*
	April	Zanzibar, then to Bombay
	July	London

When he arrived in England in 1856 Livingstone was greeted as a national hero, the Christian of outstanding fortitude, the first person to cross the African continent from coast to coast. To the Britain of his day it was an astounding feat. Livingstone was showered with honours, including one of the most prestigious degrees in the land, Doctor of Civil Law from Oxford University, and a private audience with Queen Victoria. His book, Missionary Travels and Researches in South Africa, which he wrote up rapidly from his journals, was a best seller and brought him over £10,000 in royalties, half of which he placed in a family trust.

He toured the United Kingdom speaking in support of Christianity, Commerce and Civilization. The City of London, Glasgow, Manchester, Liverpool, Halifax, Cambridge, Dublin, Dundee were among the places where he was invited to speak, and he consistently called for the 'opening' of Africa, especially along the Zambezi which was to him 'God's Highway'.

Both the Royal Geographical Society — with its support from Industry — and the British government gave solid backing for an expedition up the Zambezi, and the establishment of a research station on the Batoka plateau, above its confluence with the Kafue river. The largest industry in Britain was textiles, and mill owners wanted cotton, which Livingstone said grew abundantly in central Africa. Livingstone had also convinced his sponsors that the Zambezi was navigable with steamships as far as the Kafue, and would provide access to the interior for British traders and missionaries. During his stay in Britain three missionary ventures were set in train as a result of Livingstone's influence — he himself would subsidize a mission to the Ndebele; the LMS would establish a station in Kololo country; and the Universities' (Oxford, Cambridge, Durham, Trinity Dublin) Mission to Central Africa would go wherever Livingstone advised.

By the time the Zambezi Expedition set out in 1858, with Livingstone as leader, he had resigned from the LMS and been appointed H.M. Consul (at a salary of £500 a year) to 'Quelimane and various Rulers in the Interior', including Sekeletu. The expedition was equipped and financed by the British government, acting through the Admiralty, and apart from Livingstone, Mary his wife, and son Oswell, consisted of Norman Bedingfeld (naval officer), John Kirk (physician and botanist), Thomas Baines (artist and storekeeper), Richard Thornton (geologist), George Rae (engineer), and Charles Livingstone (moral agent and photographer). Twelve Kru seamen were recruited in Sierra Leone on the way. Mary being pregnant was left with Oswell at Cape Town so that she could go to her parents at Kuruman to have her child.

The expedition would be kept supplied by Royal Navy ships calling at the Zambezi, and was equipped with its own steamer, the Ma Robert, built specially for the purpose. It was planned to sail in her straight up the Zambezi from the mouth to Tete, and then continue up-river to the Kafue. But things went badly from the start: it proved difficult to find an access channel through the Zambezi delta to the main stream, which itself turned out difficult to navigate; the Ma Robert did not live up to its specifications, its fuel consumption was excessive, and its engine underpowered. It took months to get the whole party to Tete. To make matters worse, Livingstone and Bedingfeld quarrelled, and the latter left the party.

It was not long before Livingstone fell out with Baines and Thornton too, and dismissed them.

Then came the shock of discovering that the Kebra Bassa (Cabora Basa) gorge above Tete was impassable, a morale shattering revelation that brought Livingstone's credibility into question, for had he not declared that the river was navigable as far as the Kafue? And had he not failed to inspect the gorge on his trans-continental journey?

Undaunted, Livingstone abandoned the Batoka project and turned the expedition's attention to the Shire river and Lake Malawi. A direct consequence of this change of plan was that Livingstone was not able to give his support to the LMS mission to Sekeletu which was on its way from the Cape to the Kololo kingdom. When it arrived at the Kololo capital, the leader of the mission, Richard Helmore, and his party were received coldly by Sekeletu, who had expected Livingstone himself. Helmore and most of the party died of fever before the remnant struggled back to Kuruman. Livingstone was also unable to take the Kololo porters back home as soon as he had promised, and even then not all of them wanted to go, having settled down and married in Mocambique. When Livingstone eventually went to Sekeletu, he found the king seriously ill.

After sailing up the Shire as far as the cataracts, which he named 'Murchison's' after the president of the Royal Geographical Society, Livingstone visited the southern highlands of Malawi, and decided he had found an alternative to the Batoka plateau. When the UMCA mission, led by Bishop Mackenzie, arrived in 1861, he settled it at Magomero in the uplands above the Shire. Soon Mackenzie and other leaders of the party were dead of disease, but not before causing a storm of controversy by taking up arms against local slave raiders. In 1863 the mission was withdrawn completely by Bishop Tozer, Mackenzie's replacement as leader, and moved to Zanzibar.

Eventually the Ma Robert *sank, and another vessel, the* Pioneer, *was supplied by the British government. Livingstone had also commissioned a ship of his own, the* Lady Nyassa, *for which he paid £6000 out of his savings. The* Lady Nyassa *was intended to trade and patrol on Lake Malawi as a means of stopping the traffic in slaves, which at the time had grown to large proportions as a result of the demand for labour in Zanzibar's clove plantations, and on Réunion, the French sugar colony island in the Indian Ocean.*

Pioneer arrived in early 1861 and Lady Nyassa *(which had been shipped in sections) a year later. With the latter came Mary Livingstone, who had taken her infant, Anna Mary, and Oswell back to Scotland and then returned to Africa to join her husband. Within a few months of reaching him she was dead.*

The final stage of the expedition was a thwarted attempt to launch the Lady Nyassa *on the lake. By this time, 1863, Kirk and Charles Livingstone had left, while Thornton, who had been re-employed by Livingstone, was in his grave. In July of that year, orders came from London that the expedition was to be wound up by the end of the year. This did not leave Livingstone enough time to get his ship, which had to be dismantled and then rebuilt, up the Shire cataracts and on to the Lake. As soon as the river rose sufficiently after the November rains, Livingstone took the* Pioneer *and* Lady Nyassa, *the remains of his party and a number of freed slaves, to the mouth of the Zambezi. The expedition was over.*

It had been in the field for nearly six years, and had covered not only the Zambezi, the Shire, Lakes Shirwa and Malawi (which Livingstone claimed to be the first European to visit), but also a return journey from Tete to Sesheke, two survey trips up the Rovuma river, a voyage along the west coast of Lake Malawi, and a march from the Shire to the Luangwa watershed beyond Kasungu, and back. It had shown the extent of the slave trade in the region, and had provided a wide survey of the geography and natural resources of the areas it had covered and beyond. — Among these were coal, iron and copper, as well as potentially exportable agricultural produce such as cotton. It had also suggested a connection between malaria and mosquitoes, and demonstrated beyond doubt that that disease could usually be cured by quinine. As time would show, the Zambezi Expedition drew the map for the British conquest of central Africa.

16. INSCRIPTION

David Livingstone

Remember thy Creator in the days of thy youth. Commit thy way with the Lord — trust also in him and he shall give thee the desires of thine heart. Make thy ways acknowledge him and he shall direct thy paths.

57 Sloane St. 10th March 1857

---

17. TO H.E. STEPHENS

50 Albemarle St

11 Dec 1857

My Dear Stephens,
I am very glad to hear from you and particularly obliged by your kind invitation, but I hope to start for Portugal on the 17th and on my return I shall have very little time to spend in England.

Aye that Quilimane bar[1] was one to be remembered and I believe it hastened the insanity of poor Sekwebu. I don't know about the next trip or rather sojourn — the affair has to be mentioned in the House of Commons and then the whole thing will be in the hands of the Admiralty. I only wish I were away back to my poor friends again at Tete.

Thank your Father in my name for his kind invitation and be assured of my grateful recollection of you all. I met Peyton once in the Street in London and never could find time to call on him. I wear a cap with a gold band ever since being with you. God bless and preserve you.

I am etc
David Livingstone

1. A sand bar at the river mouth.

18. TO JAMES YOUNG

12 Kensington Palace Gardens

18 Jany 1858

My Dear Young,
I have thought much about Dechmont and the conclusions I have come to is unfavourable to my attempting it. If it should be noised abroad that I had bought an estate it would be taken hold of by some as if I had made a good thing out of being a missionary — then there is no hope of minerals as I concluded from seeing, it was just a hogs back of trap which in coming up had destroyed all the coal and I not being engaged in Paraffin manufacturing, besides the debt would hang for ever on my neck and at last smother me. A poor outlook for the Laird of Dechmont certainly. I give it up entirely and thank you for your kind thought of me anent it.

What about the engineer or engineers for I would like two, one at £150 and another at £100 or less.

Give me the dimension and dead weight of the little steam engine and sugar mill please as soon as possible. Robert's master must be paid beforehand I find. Are you coming this way again? Give up the tunnels I entreat you, by all that's good. Try to bore through Dechmont rather.

David Livingstone

19. TO A. SEDGWICK

50 Albemarle Street

6 February 1858

My dear friend,
This is the last week but one I have to spend in England and as a parting salutation I shall refer to a loving Christian letter you favoured me with more than six weeks ago. I thank you sincerely for the expressions of sympathy it contains and assure you that I go forth again cheered by feelings that I have such as you looking on and beckoning me to proceed.

That you may have a clear idea of my objectives I may state that they have something more in them than meets the eye. They are not merely exploratory, for I go with the intention of benefitting both the African and my own countrymen. I take a practical mining geologist from the School of Mines to tell us of the mineral resources of the country, then an economic botanist to give a full report of the vegetable productions — fibrous, gummy and medicinal substances together with the dye stuffs — everything which may be useful in commerce. An artist to give the scenery, a naval officer to tell of the capacity of the river communications and a moral agent to lay the Christian foundation for knowing that aim fully. All this machinery has for its ostensible object the development of African trade and the promotion of civilisation but what I tell to none but such as you in whom I have confidence is thus I hope it may result in an English colony in the healthy highlands of Central Africa — (I have told it only to the Duke of Argyll). I believe the highlands are healthy, the wild vine flourishes there — Europeans with a speedy transit to the coast would collect and transmit the produce to the sea and in the course of time, say when my head is low, free labour on the African soil might render slave labour, which is notoriously dear labour, quite unprofitable. I take my wife and one child. We are organising an iron house near the Kafue to serve as a depot that we may not appear as vagabonds in the country and may God prosper our attempts to promote the welfare of our fellow men.

With this short statement you may perceive our ulterior objects. I want you to have an idea of them. I shall always remember you at Trinity with fond affection. Pray remember me kindly and say farewell to Prof Whewell. Your auditor has given me two dozen of light ale and I hope to drink to your health and prosperity to your collegue with it on the banks of the Zambesi.

I am ever affectionately yours
David Livingstone

20. TO CUTHBERT COLLINGWOOD

18 Hart St

11 February 1858

My Dear Sir,
It is quite impossible to reply to your enquiries. I am much interested in the same subject but my facts are too few for a generalization.

I am &c
David Livingstone

21. TO RICHARD THORNTON

Steamer 'Pearl' at Sea

16 April 1858

Richard Thornton Esq.

Sir,

The main objectives of the expedition to which you are appointed Mining Geologist are, to extend the knowledge already attained of the geography and mineral and agricultural resources of eastern and Central Africa, to improve our acquaintaince with the inhabitants and to engage them to apply themselves to industrial processes and to the cultivation of their lands with a view to the production of raw material to be exported to England in return for British manufactures, and it may be hoped that by engaging the natives to occupy themselves in the development of the resources of their country a considerable advance may be made towards the extinction of the slave trade, as the natives will not be long in discovering that the former will eventually become a more certain source of profit than the latter.

2. As mining geologist you will be specially charged with the duty of collecting accurate information respecting the mineral resources of the country through which we are to travel and you are required to furnish me with reports thereon and on the geology generally of the parts visited for the information of Her Majesty's Government.

Taking for your guidance the hints furnished by Sir Roderick Murchison, you will carefully examine those parts of the country which may be pointed out to you as capable of being visited without risk to your person or health, specimens of the fossils that may be found must always be brought away as evidence of the conclusions to which you may have arrived respecting the age and relations of the different deposits, and drawings will be made by Mr Baines for the general collection of the expedition.

3. The expedition will proceed as quickly as possible through the lower portion of the Zambesi and the efforts of every member will probably be required to facilitate the transit of the luggage to Tete. While there you will have an opportunity of examining the seams of coal which crop out in the rivulets Moatize etc. a few miles north eastward of Tete and you will spare no pains in reaching a sufficient depth from the surface to enable you to form an opinion as to the quality of the mineral, where it has not been subjected to the influence of atmospheric agencies, as should serviceable coal be found at this spot (which may be considered the limit of the comparatively deep water navigation of the river) it would be far more valuable than elsewhere and its discovery will well repay the time devoted to a careful search for it.

4. Having ascended to some eligible spot beyond the confluence of the Kafue and Zambesi the iron house will be erected and experiments in agriculture will be set on foot partly with a view to promoting the health and comfort of the expedition, and still more in order to ascertain the agricultural capabilities of the country.

The Zambezi Expedition

Whatever knowledge of soils you may possess will now prove of great value and you are expected to communicate it freely for our guidance.

5. The iron house being established as a central depot, explorations will be made in various directions in company with the Makololo, and botanical and mineral and zoological collections will be deposited at the central station. But though these explorations and collections are very desirable you will understand that Her Majesty's Government attach more importance to the moral influence which may be exerted on the minds of the natives by a well regulated and orderly household of Europeans setting an example of consistent moral conduct to all who may congregate around the settlement, treating the people with kindness and relieving their wants, leading them to make experiments in agriculture, explaining to them the more simple arts, imparting to them religious instruction as far as they are capable of receiving it, and indicating peace and goodwill.

6. It is hoped that no action will ever arise in which it will be necessary to use our firearms for protection against the natives but the best security from attack consists in so acting as not to deserve it and letting the natives see that you are well prepared to meet it. You are strictly enjoined to exercise the greatest forebearance towards the people, and while retaining proper firmness in the event of any misunderstanding to endeavour to conciliate as far as possibly can be admitted with safety to our party.

7. Your own principles will lead you in all your dealings with the people to follow the strictest justice but it is necessary to remind you that even the appearance of over reaching yourself and insulting must be carefully avoided. No native must be employed unless a distinct understanding has been come to in the presence of witnesses as to the amount of remuneration to be given. It will greatly facilitate your intercourse with the people in the middle of the country if you acquire the Sichuana language.

8. You are distinctly to understand that your services are engaged for two years unless any unforeseen accident should happen to the expedition when you will be set free as soon as an opportunity is afforded for returning to England.

9. I hand you the appendices number one and two for your information in order that should any useful plant or new animal come under your observation while engaged in your other duties you may communicate with the head of the expedition. While it will be necessary to employ our firearms to procure supplies of food and in order to secure specimens of animals and birds for the purposes of natural history the wanton waste of animal life must be carefully avoided and in no case must a beast be put to death unless some good need is to be answered thereby.

10. Finally you are strictly enjoined to take the greatest care of your health — avoid all exposure to night exhalations and should you be troubled with drowsiness constipation or shivering apply promptly to Dr Kirk for medical advice. Trusting that our heavenly Father will watch over you and that you will return to your friends after having performed your duty with honour I heartily commit you to the care of overruling Providence.

I am your most obedient servant
David Livingstone

52

22. TO SIR MORTON PETO

Zambesi River

21st June 1858

Dear Sir Morton

We have been for more than a month exploring the delta of this river in order to find a safe entrance, and having succeeded beyond our expectations after giving up hopes of entering by the Luabo as recommended by Captain Parker we find it prudent to send away the Pearl and go up by the steam launch Ma Robert. The Pearl draws 9 ft 7 and though at this time of a falling river we could get a channel never less than 12 ft as we were specially charged not to risk her detention in the river we have set up the iron house on an island about 40 miles from the sea and when all our goods are in it the Pearl will be sent off to Ceylon. We shall take them up to Tete by successive trips and the iron house last of all. Now if we had a sloop, (paddle) of four or five feet such as you offered to build we could even now go with ease up to Tete. I gave your estimate to Captain Washington and he took a copy and desired me to apply to the Foreign Office for such a vessel and he will I am sure second the application. But these matters are generally put into Macgregor Laird's hands in this way. He writes to Washington an official letter telling what kind of vessel he thinks best for such an expedition and as it saves the great folks the trouble of thinking the job is at once handed over to him — By this mail I apply for the vessel described by you and I earnestly hope it will be put into your hands for I am sure you will do the work efficiently and *con amore*[1]. I don't know whether you are acquainted with Washington if so you would hear how my proposition is received and might assist me otherwise by your influence with the higher officials.

The first news we got of the Portuguese was that they had been expelled the country by the natives — this alters the aspect of affairs in the river for me — but as it happened before we came we cannot be blamed for what the Portuguese call rebellion. I visited a party of the rebels at Mazaro and though about 200 appeared well armed ready to dispute our progress on my calling out that we were English they raised a shout of joy and at once ran off to bring bananas and fowls for sale. We have had no fever — Mrs L was obliged to leave us at the Cape but fortunately met Mr and Mrs Moffat there and will after her confinement join us by coming overland.

Sorry I did not get Havelock's life before leaving. My kindest salutations to Lady Peto, Aunt Helen and Mr Brock.

David Livingstone

Margin   I have referred to your offer in my dispatch and told Lord Malmesbury that a copy is with Washington.

Margin   We have low reaches of four or six miles of from 5 to 7 fathoms then a crossing of 12 ft but it has to be searched for.

1. *Italian*: with love.

23. TO RICHARD THORNTON

Island of Pita

25 August 1858

My Dear Mr Thornton,

We have found the Pinnace[1] too heavy for the engine in low water where we are obliged to go slowly and cautiously and I have it at Pita to make the first trip in the launch alone. Mr Baines volunteered to remain or go down to Shupanga with the Pinnace but I would not hear of it at first. He has however improved so much and we are the [?] country that I consented to his remaining in charge close to a village belonging to the Portuguese of Senna. This will make you exercise your patience longer than we expected but if I find the water too shoal [sic] for the Pinnace to go easily I may come down at the time we thought after depositing the cargo of the Pinnace at Senna. We could not well leave it at Senna as we cannot come near the village and the carriers urged on by a yellow halfcast are disposed to be extortionate. Unfortunately the Major's letter came back so we must look for the coals ourselves or wait three days till his people do. I hope you are enjoying yourselves in botany and geology. In making your copies of the chart put in native names or those of your own invention rather than Bedingfeld's. To you belongs the honour of first laying down those portions of the river and your names ought to stand. Thus what he would call Arab point from seeing an *East Indian* there call Nyuruka point. Sunday island or [?] island call by something in their geological formation to make your own stamp on them. I missed your chart badly.

As I am required by the Government to furnish the Secretary of State for foreign affairs with a full statement of the case of Bedingfeld I shall require from you for the use of that minister your evidence as to what you heard when I requested B^d to adopt some system in the expenditure of the Kroomen's provisions — that 'he would have nothing to do with it' — 'that I ought to have got a lighterman instead of a man of his standing' etc or words to that effect, and as it is on such evidence as yours that the Foreign Secretary places great reliance you might state any opinion you may have formed as whether the Expedition could have gone on efficiently in the state of insubordination into which Bedingfeld was drawing it. This however is optional. The evidence is to be written on foolscap paper and the original alone sent. It is not made public but for the use of the Foreign Office alone.

If you can remember that B told me he had written nothing up to that time for the public perusal please to do so. I shall send off a packet when I come down. Your charts will go to the Foreign Office and thence to the Admiralty. Any packet of private letters sealed up with the word(s) only (Letters for England) will if you choose be opened in the F. Office and stamped and distributed by the Post Office free of charge to your friends.

God bless and keep you. Keep in company as much as possible with Dr Kirk while geologizing — for the sake of safety.

I am etc
David Livingstone

1. Small light boat.

24. TO SIR MORTON PETO

Ma Robert
Zambesi River

17 Septr 1858

My Dear Sir Morton
You may remember that I had the offer of a vessel built by Mr Scott Russell called the Bann and that on a report of Commander Bedingfeld I was led to reject her — I have since found out that Bedingfeld made a false report both of the Bann and of this vessel the Ma Robert. He was exceedingly anxious to get a sailing master but on my declining to apply for one he went and made some enquiries at a low carpenter about Woolwich and trusting to B's statements I rejected what I now find would have navigated this river the whole year for she draws only three feet. Bedingfeld suffers so much from seasickness and it grows upon him that he could not as was expected of him have navigated the Bann out. Bedingfeld then when sent to witness the trial of this vessel on wood alone reported "that she was just what we required and held her steam well on a very little wood." Trusting in him again I was undeceived only when on the departure of the Pearl we had no more coals. The engine is badly constructed — has the boiler tubes below and on one side the fire — takes between four and five hours to get up steam and we must cut a day and a half to get one day's work out of her — this though we use lignum vitae[1] which here abounds and when dry is excellent fuel — the canoes pass us and look back at the "Asthmatic" as she ought to be called and I bitterly repent rejecting the Bann, for having been up at Tete and this being the lowest state of the river I can now give a positive opinion — I don't know whether Govt will give me another steamer after rejecting the other but I shall do the best I can with this. Bedingfeld turned out a most cantankerous subject otherwise, so I send him home where he will abuse me to the utmost. His powers of assertion are truly wonderful and being more than ordinarily ostentatious in his secret devotions it has been a great affliction to have had anything to do with him. I sent home to Government his first report of the capabilities of the Zambesi. It was very flattering — now that I have been forced by his overbearing conduct to the other officers to dismiss him, he will give a report the very reverse. He thought himself indispensable as there was no sailor among us

and gave in his resignation several times — I took it at last — mounted the paddle boat myself and made quicker passages in conveying our luggage from stage to stage than he ever did. At Tete we were the first steamer they had ever seen and five Portuguese gentlemen at once set about getting a supply of coals. In three days we got a ton and a half — the Engineer thinks them very good though all from the surface where they have been exposed to the action of the air and water for ages. He says when we get them a few feet in he has no doubt they will equal the very best Scotch coals. This is the beginning of the coal trade here. We have no end of the best of specular iron ore. With coal and iron surely we shall do something to open up Africa — the rapid above Tete is I am informed not a waterfall but a series of rocks jutting out of the stream, if they can be blasted, as there is a deep channel among them don't be surprised though in addition to turning skipper I try to be a quarry man.

The people are at war with the Portuguese and we pass and repass from the one side to the other simply because we are English. I landed once unintentionally on the battlefield — the sight of the dead sickened me and the Governor being very ill of fever his officers begged me to take him on board. The firing was resumed about as far off as the opposite side of the street from your house and all was confusion. I sent Bedingfeld to bring our men to carry the Governor — the fellow tried instead to get the steamer underweigh, but the Engineer refused to set her agoing — seeing no one coming and not relishing the idea of a ball boring a hole in my precious body I took the Governor up and carried him off and cured his fever.

David Livingstone

My kindest regards to Lady Peto and all the family.

Margins The felt is a mistake for a warm climate — an iron roof would have been better. We had the house up once.

1. A tropical hardwood, *combretum imberbe*.

25. TO CAPT DENMAN

Tette

19th February 1859

My Dear Captain Denman,
We examined the rapid above this at low water and found it a great curiosity for instead of a rapid it was a deep groove in a rocky bottom and as we steamed along the man at the lead kept calling 'no bottom at ten fathoms'. The breadth was from

40 to 80 feet above us. In this groove there were some cataracts but when the river is in flood all disappear. I sent Mr C Livingstone up to examine it when the river had risen 12 ft. He found all smooth except one cataract though the river had fallen several feet by the time he reached it. But we need a strong steamer to go up — this with [?] all her steam on [?] stands when the current is three and a half knots — At the rapids it is 5 in the most rapid parts & in the rest 3 3/4. I felt inclined to drag this up but feared a collapse as she is only 1/16 of an inch in thickness and even if we had succeeded we should soon be out of supplies after we had passed for she can carry but little cargo —

We therefore work for the present below it and our first trip was up the Shire a branch which as far as we can learn has never been explored before. We found a good navigable river for at least one hundred miles from the confluence — the people were very much alarmed and watched us night and day well armed with bows and poisoned arrows. The valley of the Shire is about 20 miles wide — and bounded by lofty mountains — the upper third of it is well cultivated — On a large swampy plain we passed many hundred elephants — fine fellows with enormous tusks. We could chase them sometimes for there are branches which depart and re-enter and make islands and these branches are usually deep. A cataract stopped our progress but we heard that five days land journey led to the river smooth again and used by the Arabs to come down in canoes from Lake Nyanja. The people were so suspicious that we thought it would be imprudent to leave the vessel among them for a land journey but we return again in a month or so and may obtain the confidence of the natives.

This is our sickly season but though we have fever here none die of it as at the coast, so say the Portuguese. We have had three mild cases and they are well. This is the edge of the interior healthy highlands — I have enjoyed continuous good health ever since we came — Thanks to Him who has protected us we are all [live] and well. We want to see if [they] will send out the Bann or another strong vessel of light draught to take us and our luggage up at once; the Bann was given to us — Bedingfeld declared himself perfectly satisfied with her but he wanted a sailing master I refused and he refused the Bann. I trusted implicitly [in] him but had misgivings when I saw him seasick. We have applied for her again but I fear it will go down after declining her once. If you could say a word for us anywhere I know you will do it. I have guided this thing 1600 miles — Half our time is wasted cutting wood — we have cut upwards of 100 tons of lignum vitae alone.

Ever yours
David Livingstone

26. TO W.C. OSWELL

Chibisa's vil.
River Shire
Lat. 16° 2 S. Long. 35° E

2 April 1859

My Dear Friend

A year has nearly elapsed since we parted, but I have been unable to write you a line — My naval officer thought we could move neither hand nor foot without him and resigned first, when we entered the Zambesi, and subsequently when he thought I could not get rid of him. I was as mild as possible till I saw that he meant to ride rough shod over all authority — but then I assumed the charge of the steamer, and when I made the first trip more successfully than he had ever done I never saw human face lengthen as did Bedingfeld's — When I gave him notice to quit he began to dance and sing, but when he saw we got on better without him than with him, he tried to induce the other members to remonstrate with me. They gave me written declarations instead that I had acted quite right so off he was obliged to go. Not however before doing us an immensity of harm — He overloaded the expedition with baggage — the Government gave me a fine steamer — the Bann — and as it drew but 3 feet it would have suited admirably. Well he applied to me for a sailing master and having refused on the ground of wishing to have but few Europeans he re-examined the Bann and though at first he said she was all we needed, she was now unfit for anything — I of course yielded to the opinion of my "naval officer" and she was rejected. When we got on board ship we saw that B could not have navigated her out from sea sickness! I never quarrelled with him but he quarrelled with the Captain of the Pearl and others and it is a relief to have done with him though from my having turned skipper myself I have not been able to write to my friends. Then the vessel we have — the Ma Robert is such an awful botch of a job — I have been obliged to take up our luggage and provisions in driblets having passed over 1700 miles at least and we have to cut a ton of wood for 7 or 8 hours steam — a current of three and one quarter knots (patent log) holds her back so that she cannot gain an inch and so does a stiff breeze. I have applied to Government again for the Bann and if she is not granted then I have ordered £2000 to be expended for a vessel of my own. We could have passed up the rapid of Kebra basa in Febry last had the vessel possessed any power but she has none for the current then flowing and being only 1/16 of an inch thick we were afraid she would double up had we attempted to tow her through by means of the Makololo.

I may tell you that between ourselves Bedingfeld was an awful bore from extra ostentatious piety — associated with a terrible forgetfulness of statement. His *private* devotions needs must be performed in the most public place in the vessel — and he tampered with the Kroomen — telling them he would soon be out again to Sierra Leone in command of another ship and would give jobs to those he knew etc. — but they nobly declined to be led into mutiny by him. He told this himself — Proposed to the engineer also "to lay our heads together and then we can do as we like". If you have a war with such fellows to command woe betide our navy. Some Lord or other pushes him on in it.

As we are below the rapid and our establishment is at Tette we are examining the country adjacent. The Shire promised best but we went up a hundred miles of latitude from its source in January last and found it a fine river for a steamer. It really does come out of the Lake Nyanja but a little above this its navigation is obstructed by cataracts of the same rock and nearly in the same latitude as Kebra basa — the people Manganja were much alarmed at our presence so we thought it unsafe to leave the vessel and go overland but returning to Tette we remained there during the most unhealthy month and have now come back to see what can be done. In the interval of palaver about guides etc. I write this to you.

There is a high mountain near its entrance called Marambala. It is 4000 feet high and a fine point for a healthy station. It is well cultivated on the top having hills and dales and flowing fountains there — Lemon and orange trees grow wild so do pine apples. Above Marambala the country or valley of the Shire is marshy and the river winds but not so much as the Chobe. Above this marsh there is a large plain and prodigious numbers of elephants were seen both times on it — there are branches forming islands and when we got the elephants on these we hunted them in this steamer — With a good one we could have killed the whole herd. I saw five or six herds of them at once, my companions estimated them at 3000 (I think 800 or less) but one herd had fine tusks — probably all the males had, the bodies are not large. There is plenty of game in the country but there is no hunting it — the grass is so tall one is lost and cannot get well along except on paths — I keep a corner to tell you how we get on

We hope to start on the 4th.

Shupanga 14th May 1859. We left on the 4th and after a fortnight through a well peopled mountainous country came in sight of a high mountain called D Z O M B A or Zomba in the East we are now 1500 feet higher than the vessel and the Shire seemed coming round the North end of Zomba, but a marsh prevented our going that way — after we tried it we went South a little and crossed the spur of the mountain and first got a sight of Lake Shirwa on the 14th April; on 18th we were on its shores and a magnificent sight it is for it is surrounded with lofty mountains. Ngami is a mere pond compared to it. We went up the side of a hill but could see no horizon but water in the NNE and 26° of that — only, two mountain tops rose in the blue distance like islands 50 or 60 miles away. It is 20 or 30 miles broad of a pear shape but the tail extends some 30 miles South of the body of the tadpole. As you know me to have been always a dead hand at drawing look here and admire!

[There is a sketch at this point in the original.]

There is no outlet known — the water is bitter — a little like a very weak solution of epsom salts. We think it is 60 miles long exclusive of the tail and the people told us that it is separated from the Lake Nyinyesi by a narrow neck of land — We slept at the village marked 0 and they compared the distance across from Shire to Nyinyesi to what we came that morning = 5 or 6 miles — Zomba or as it was sometimes pronounced Dzomba being over 6000 feet to a person in Nyassa or Nyanja it will appear in the Lake as one of the church missionaries described it — we patronize Nyinyesi — the stars — or Nyanja by which they sometimes call it means any large water. Well we went down the Shirwa valley and in crossing over the range which separates it from the Shire valley we got a glimpse of the end of the tail — Went down the Shire till we came to a branch called Ruo which rises in that high range called Milanje — ascended it 7 or 8 miles and found we were about 30 miles of Lat from Shirwa. If we don't get another vessel soon we go up the Ruo — carry a whaler over to Shirwa then go to Nyinyesi — this is for yourself only.

We could not hear a word of Burton — I have had no news from home — if he has discovered Nyassa first, thanks to my naval donkey — I would have done all this last year — but we have got a lake of our own and a short cut to his. Dr Kirk and 15 Makololo formed our party — country high and cold. No game but on the West of the Shire — the Maravi country — there is plenty — all we met were Manganja and one party of slave trading Bajawa. The Portuguese knew nothing of Shirwa. It is 15° 23' South Lat 35° 35' E Long. On coming down Shire we killed two elephants and wounded several — caught a young one but one of the Makololo in excitement cut his proboscis so that he bled too much for our medical skill he died after two days. The Portuguese are bent on shutting up what we open — they never went up the Shire and dare not now — an old lady at Tette distinctly remembers two black men coming to Tette and they never went farther — consequently the claim set up for them fails for 300 miles.

David Livingstone

27. TO JAMES YOUNG

[Kongone?]

[1859]

Additional

My brother and I have concluded that it will be right to try and get up a trade in the produce of this country as we despair of the Portuguese ever doing anything except in ivory and gold dust. We are not men of business — witness the way Macgregor Laird Jewed us after pathetically advising us "not to take a gun boat". McCurndle Shaw & Co, Glasgow, know the proper things for African trade. The things must be strong generally not flimsy dirt though some of them are not such as can be used in England they will do for some things here. A portion such as may be given or sold to chiefs may be in the form of "Panno da Costa".[1] These panno da costa things are best for trade and so are imitation African cloths.

We send £200 — half of which there is an order for my brother's agent and the other half you must pay yourself from the money you may have in hand. I am supposing that another settlement may have been made with Murray. I would have sent an order on Coutts at once but this goes through Portuguese hands three or four times and I don't trust them much.

My brother's agent is Good & Co, 16 or 21 Surrey St, Strand. If you don't get an order in a letter from him which accompanies this please stop payment there. You are not bound to McCrundle Shaw & Co but they sent us a present of very proper goods for this trade and can serve us well if they liked. Excuse our presumption in trying to employ a landed proprietor for the benefit of our peddling. Had we possessed a proper ship we could have paid off the whole by lignum vitae and ebony. We are not sure of succeeding but try, though trade for me at least is against the grain. I think a bale of good blankets white and coloured might be tried too. We must know the prices too, exactly, in order to sell fairly.

Had we English here they would get all the trade. The Germans will probably do something.

We are struggling up stream. I put Rae to navigate yesterday. He said he never had any idea how tedious it was. We beached her at Kongone. The steel plates exfoliate and leave a bright new surface and are now so thin they cannot be mended. A little tap with a hammer would go through — we puddle[2] her with clay instead. I used Mr Scott Russell very ill in taking the opinion of a naval donkey instead of his. I wrote him but he cannot forgive me I fear.

Every blessing on you and yours
David Livingstone

1. *Portuguese*: cloth of the coast. A type of calico.
2. To render water-tight by lining with clay.

28. TO JOHN KIRK

Dakanamoio River Shire

17 Octr 1859

Dr Kirk

Sir

You are hereby required to pass overland with Mr Rae to Tette in order to bring away two persons lately members of this Expedition in order to send them home by the Man of War appointed to meet us at Kongone Harbour in the middle of November next.

As Mr Thornton, one of the persons referred to has been honest, and failed in his duties as geologist chiefly from ignorance and a want of energy, he is permitted to take the geological specimens with him but on the understanding that they are Government property and must be handed over to the Geological Society when required. He must give you an acknowledgement in writing to that effect, otherwise, they are to be retained at Tette.

The other individual, Mr Baines referred to having been guilty of gross breaches of trust in *secretly* making away with large quantities of public property, and having been in the habit of secreting Expedition property in his private boxes, it will be necessary for you to examine his boxes (after ascertaining whether my order to him to deliver up to Major Secard all paintings, drawings and other public property has been complied with) the more especially as he had only three *private* boxes in his possession on leaving the Pearl, and these are now increased by the appropriation of "biscuit boxes" to which he has no right — no permission having ever been asked or granted.

It will be proper for you to ask him in the presence of Mr Rae — what he did with five jars of butter which he took out of a cask and never sent to table or for cooking. What he did with five barrels of loaf sugar which he was seen opening and drying but were never used in the Expedition. The answers to these and other questions to be put down in the storekeeper's book as soon as convenient and signed by yourself and Mr Rae. Take possession of this book — of another book of mine in his boxes "The Plant", of specimens of brass rings and of everything else you have reason to believe does not belong to him. If he declines your offer of conveyance he is left to his own resources.

I am etc.

David Livingstone

29. TO W.C. OSWELL

River Shire

1st Novr 1859

My Dear Mr Oswell,

We have travd this river up to its origin in Lake Nyasa or Nyinyesi in South Lat 14°
25'. We go (steam) up 100 miles or so from the confluence, then pass 33 miles of
cataracts beginning at 15° 55' then the Shire is smooth again right into the Lake
Shirwa which simply means "big water" is called T A M A N D U A, we found it to
be about ninety miles long, but no one could tell us how far off the end of Nyassa
lay. The natives think that it goes a long way to the North "then turns round into
the sea". It had a heavy swell on it though there was no wind and it never rises or
falls much that we could notice, and gives of Shire constantly, i.e. a river of from 80
to 150 yards wide or 12 feet deep with a two and a half knot current, and never
varies more than 2 or 3 feet from the wet to the dry season — it must be a large
body itself. But we could not explore it for we had left Mr Macgregor Laird's
precious punt in a sinking condition and had to hasten back: Furnace Deck and
bottom all became honeycombed simultaneously that was £1200 (not including
extras) for 12 months work — Fairish rather, without any whine of *"doing it all for
the good of the cause"*. Well we found that there is only a small partition between
Nyassa and Tamandua; but we could not examine it as we were on foot, and had
been longer away then we promised. Along this partition all the slave trade and
other trade must pass in order to get away to Mosambique and the other ports on
the East coast. We met a large party with an immense number of slaves and
elephants tusks, coming from Cazembe's country: and bought some fine specimens of Malachite from them — they were not Arabs but looked somewhat
like them — an awfully blackguard looking lot. When they knew we were
English — and saw our Revolvers they slipped off by night — probably thinking
the same of us as we did of them — An English mercantile [?] would eat out the
slave trade, for the Babisa and other native traders, would not go a month farther
(to E coast) for the same prices we could give here. I propose to Government to
place a small steamer on Nyassa — a common road could easily be constructed
past the cataracts and the vessel made to be unscrewed and carried overland.
We are going ahead in our ideas you see, but it's your own idea suggested
long ago in the Kalahari that I am bent on carrying out. Look at it and see if I am
visionary. Above or say abreast of the cataracts, the land East of Shire is in three
terraces — the lower or valley of the river, strongly resembles that of the Nile at
Cairo. This one is 1200 feet high — the next is over 2000 feet and 3 or 4 miles
broad — then the third is over 3000 or about as high as Table mountain.
I never saw any part better supplied with rills of running water than these terraces, and my companions Dr Kirk, C. Livingstone and Mr Rae declare the
same thing. Cotton is cultivated extensively over them all. On the last terrace
which is some 12 or more miles broad and sloped down to Tamandua, rises Mount

Zomba, which we ascended and found to be in round numbers between 7000 and 8000 feet in altitude — We travelled in the hottest period of the year — that immediately preceding the rains — Shire valley was hot and stifling, but no sooner did we ascend to the second terrace, than the air had a feeling of freshness. On the 3000 feet terrace it was delightfully cool, and on Zomba it was cold. We have thus changes of climate within a few miles of each other: Europeans without doubt could live there: the people have no cattle, but cultivate largely, and the trade in cotton could be developed I think by a small colony of our own [letter torn] this would be a mission [letter torn] us of our dependance on slave labour. This might be attained after our heads were low but the field watered by the Lake and its feeders may be called a cotton country of unlimited extent, which really seems superior to the American, for, here we have no frosts to cut off the crops — and instead of the unmerciful toil required in the slave states, one sowing of foreign (probably American) seed already introduced by the people themselves serves for crops for three years, even though the plants should be burned down annually. There may be evils to counter-balance these advantages but I don't know them yet. The people are said by the Portuguese to be of quick apprehension. We removed their suspicions that we might be a marauding party by frankly telling them that we came to mark the paths, for our traders to come and buy cotton. We go up to the Makololo country in about three months — We have been turned aside for a time, but do not regret it — Nothing can be done with the Portuguese — they are an utterly effete, worn out, used up, syphlitic race: their establishments are not colonies, but very small penal settlements.

We can cure fever readily. The only one likely to be cut off is our miserable artist. I left him at Tette by way of sparing him exposures to malaria in the lower Shire and he took to stealing our stores and drinking and debauchery [letter torn]

---

30. TO JAMES YOUNG

Zambezi

28th January 1860

My Dear Young,

A mail bag was lost last month by the capsizing of a boat from MS Lynx but happily no lives were lost — we sent a mail by that vessel but neglected to tell you that the loss has entailed on us ignorance of almost every private affair. We have not for instance the smallest idea whether we are ever to get another steamer though our ideas vegetate towards two of them. In a sort of despair I sent Rae home. His time is out and bamboozled by the wretched conduct of Mcgregor Laird,

Bedingfeld & Co we have no work for him. He will be able to benefit us more by going home and working for us there than he could do here. If you find from Captain Washington that the Government will not give a second steamer then I wish Rae to see a small one being made. We have talked the matter over and agreed as to the dimensions. This is for Nyassa and is to be capable of being unscrewed at the foot of the falls in the Shire. It must sail out or steam out and not be carried. Perhaps I am boring you, and annoying you with my affairs, but I trust in you still. My idea is that we may do an immense work by a small missionary and mercantile establishment up the Shire. I have proposed it to Lord John Russell but don't know how he is inclined to it. I am determined to go it through if he does not. I shall pay for the steamer — £2000 or £3000 and the persons who may be got to engage in the enterprise may supply provisions and seamen's pay and expenses out. If I could speak more pointedly I would do it.

Mr Braithwaite the solicitor seems interested in the matter and talks of it not seeming difficult to raise a vessel among friends. Will you talk the matter over with him? To me the matter appears in this way. If the Government does its duty — well — if not, I shall do it myself. The thing must be done. Rae is instructed to tell you all he can. I have given you as full information on every point as I could, but may have said too little about the Portuguese. Anyone coming out here must give up all idea of co-operation or aid from them. They are all the lowest of the low — there is not half a dozen upright men of that nation in Africa. Their establishments are small penal settlements and not colonies. No women are sent out, and the moral atmosphere is worse than that of the valley of Siddim. Some have been very kind to me and I would not like to pain them by saying much, but a more effete syphilitic brood does not exist. It is the utter worthlessnes of these Portuguese that renders a vessel such as I applied for first i.e. one capable of going from the lower Zambesi to the Cape, necessary, as well as a small one for the Lakes. For both, lightness of draught are desirable. In the name of all that is good, don't kill us with wood cutting. Here we are cutting a ton of dry lignum vitae for every five or six hours steam. Let us have a condenser. The wood cutting has been a heartbreaking affair. Furnaces fit for burning wood ought surely to have been given in the present case, but everything furnished by the Laird's seems to have the superscription "They are going to the dogs like the Great Niger Expedition and anything will serve."

I try to forgive those Birkenhead Jews, but feel like Rob Roy "Well I forgive but my curse on my sons Rory and Murdock if you ever forgive". I become very much annoyed when I feel compelled to do work imposed on us by fraud instead of good service to the cause of African elevation. But all will come out brawly[1] yet. "There's a good time coming yet, a good time coming".

I sent by a Lieutenant Medlycott two watches for repair. One is of the only really gentlemanly Portuguese in the country Sr Ferrao. I beg to trouble you to give them to Dent in Strand. When sent out let them be insured. We were cut off from communicating with the ship or I should have written about them. He was either to send them to you or direct to Dent. If gold spectacles I ordered were sent they must now be at the bottom in the mail bag. Knapp's Technology [and] Murray's statement at the bottom too. Any of my friends who may have written must write again.

We get "Saturday Reviews" awfully bitter — enough to give one bilious attacks. We should like Chambers' Journal better, but Dr Kirk will let you know officially.

I got a letter from my Canadian brother, he seems likely to do well under your kind patronage and both he and I are thankful for it. Charles gets on pretty well, but new chemicals are necessary for his photography.

I told you of Baines & Thornton — one can't make silk purses out of sow's ears. Missionaries even often turn out so and so, and what can we expect?

I don't know anything about my family or John Moffat except what I glean from the public papers. I suppose you have been to Aberdeen. You always go but don't read papers the more shame to you and some other delinquents I could name.

The scraps we have got about Burton and Speke are in favour of my view of the continent much beyond where I travelled. This is gratifying.

The Zambezi was again exceedingly low this year but has begun to rise, and we shall probably have a succession of high years. We are beginning to wonder if the other end of the Zambezi has any connection with Nyassa. It is a great affliction to have been prevented so long from going up to the Makololo, but we go now either afloat or afoot. We cannot however take up our sugar mill, engine etc. without a steamer.

We look for guidance from a High Power and have not been disappointed at that source yet, so will persevere. Is Murray going to send out a cheap edition? I have heard nothing from him about anything, but have received three Quarterlies.

Any heavy thing would probably come better by Captain Washington. We shall be away some 6 months at least, but possibly may write by way of Kuruman.

I report a heavy bill for travelling expenses of Ma Robert[2] but as I mentioned her letters were lost. My love to any lady you may see with a good sousy double chin and to all the family.

Ever affectionately yours
David Livingstone

Margins:
1. Will you send me a book by Bishop Selwyn? It contains all the words in the Bible. It is among those I left at home.
2. Will you enter into communication with Mr Braithwaite about the steamer? His address is 3 New Square, Lincoln's Inn, London.
3. He speaks as if something could be done. There is a mission getting up at Cambridge. If you get on with the steamer let them know. I must trouble you so excuse me.

1. *Scots*: splendidly.
2. i.e. Mary Livingstone.

31. TO JAMES YOUNG

[7 February 1860?]

A trader called Chapman has published a book on his travels. His partner Samuel Edwards a missionary's son stole a volume of my journal. At least the Makololo who know him well said we "gave it to Samo". He denies having got it so: I am curious to know whether his partner has made use of it. It contains many notes on wild animals got from Bushmen. Send me a copy by Washington.

We hear that Bedingfeld has been shewing a letter written to him by Captain Gordon, to everyone who will read it. Gordon tries to blacken Captain [Duncan ?] and me to the utmost and in doing so asserts things on Bedingfeld's authority of which he has absolutely no personal knowledge. I am rather uneasy for England since I became more intimately acquainted with some of her defenders. Ask Rae about Bedingfeld's performance on the battlefield at Mazaro. It was a regular case of shewing the white feather. He was quite pale and so frightened he forgot all about my order to send the Kroomen.

I am extremely anxious about the other steamer for the Zambezi. The want of it brings all our contemplated operations to a stand. We cannot go up to the Makololo country except on foot. This will be immense toil and no gain for we cannot take sugar mill or anything. It would be a flying visit only and next month or April ends our chance of going up the rapids of Kebra basa. Our next move suppose the vessel does not come in time is to try the Rovuma, which we believe comes out of Nyassa and enters the sea beyond Portuguese territory.

In the meantime we try to get the Portuguese to consent to freedom of access by Zambezi. We have doubts yet hope in Lord John Russell. If he does not get it then we must go in by Rovuma. Macgregor Laird and Bedingfeld, they have done us no end of harm. I hope you do not feel bored by my asking you to do so much. No letters from you except one since I left home in April 1858.

Between the Penny Post nuisance at home and the no post abroad it's enough to make one take the megrims. I don't know anything about my wife but trust it's all right.

Did you ever try to get rid of the smell of your oil by putting a single drop of turpentine in it or putting it in a bottle in which turpentine had been? I have found both odour and colour of iodine solution disappear from being put into [a] turpentine bottle which had been only slightly washed though the colour as deep previously as port wine. Don't forget to send us some oil with Rae's vessel and your stunning lamps too of course. I suppose you have improved them so much that you can shew a needle in a bottle of hay. They will do more good than the famous Tunnel that once disturbed your dreams, I say. Get a brooch and pair of bracelets made out of malachite for my friend that sits at the head of your table. I don't know her name yet. Baines has taught me how easy it is to make presents of what does not belong to one.

Present kindest salutations to all and may the Good Lord dwell in your hearts by love.

Ever Affectionately
David Livingstone

32. TO JAMES YOUNG

HMS Pioneer
in Zambezi

14 May 1861

My Dear Young,

I find that I am in want [of] a new blue frock coat and am afraid that this may not reach you in time for one to be sent out with Mrs L. I want it made in no particular fashion i.e. it must not be made in the fashion of the day as it must serve me several years — buttons of the consular uniform. Messrs Brownlow & Oliver have my measure. I think they live near Regents Circus but a friend of mine Mr Starkie of New Bond St [in] the London Directory can give the address. I don't like it to hang on flappingly.

If you have sent out the Nyassa steamer the arrangements will all have been made but should any delay have occurred be sure to send pointed instructions as to the delivery over to me if that is the plan, and trust nothing to the good intentions of those who may have her in charge. I suppose that the Government may have taken her off your hands, though you proposed to send her out as a private gift — for they do not like to have an enterprise of a hybrid character. If built by Tod it must be a good one. And if drawing little, say less than three feet, she will serve us admirably. We here draw five feet though intended to draw only three. They ought, one would think, to try for a draught of one foot when they mean to get one of three.

We hope soon to get up to the Lake. We came up to Mazaro some 70 miles in a day and a half or in 15 hours — last time we came in the Ma Robert the same distance it took us ten days. Accidents have kept [us] back a little since but we shall soon go up Shire and land the mission.

Rae behaved very well when with us but you must not trust him. He did some queer things and his application to the Government for £300 as a valuation of his losses is outrageous. He had not fifteen shillings worth of clothes with him when he left us, and he abandoned the specimens with which he was sent in charge. He had no luggage when in the Lyra; and if he speculated in gold dust at Quilimane the loss could not be charged to Government. I have accidentally discovered some of his very mean double dealings with myself and although I mean to take no notice of it I shall certainly be on my guard.

In that unfortunate affair with my brother John I don't feel easy at not having

68

given you an opinion though by sending the document[s] you evidently intended me to form one. This I assure you was only because I did not wish to run the risk of endangering our friendship by writing, the tone of which might be read very differently from what was intended, but silence again does not seem respectful especially to [one] whom I am sure wishes to do only what is just and right. (I avoided also giving any opinion to my brother or to any one else).

There can be no doubt but you are legally right. That is I think that you will have the law on your side, because you had a perfect right to sell the goods at a higher rate to him (as he believed), the sole importer into Canada than to customers at home. He spent all his little property in advertizing and would have come out well it seems had not umbrage been taken by your partners or self at the little he had, I suppose, without leave assumed.

Then another party or parties in Canada were supplied with oil at so much a lower rate that they actually offered to supply him more cheaply than he could get it from you. When I say *you* I of course mean your firm. He might be misled by the Canadian parties making the offer, but he had the statement of one of your own agents at home that a larger price was demanded from him than from the others. My opinion then, in which I may be quite mistaken, is this. He relied too trustingly in the idea that by your friend[ship] he had secured not only the first but the sole right of importation for a time at least. This did not comport with your plan. In fact he had no right to assume so much and he was let down by other agents being granted the privilege of importation and at a lower rate.

In my own mind I explain this by supposing the lower rate to be been [*sic*] granted by another member of [the] firm without your knowledge. If your books show this, and no one else has a right to ask whether they do or not, I hope it may modify your law proceedings against him. I believe him incapable of fraud. He has a very high character for integrity in Upper Canada where he is known.

I feel that I led him into this unfortunate affair, you too thought it would be advantageous for both and must regret that it has turned out so differently from what we expected. All I presume to ask is to reconsider the case in view of the poor fellow being "nursed" by other agents after he had spent his all in advertisements, and don't push your legal proceedings too hotly. If you take a totally different view of the case from what I do pray do not trouble yourself to reply on the subject but let us continue our correspondence as heretofore.

We have had a good deal of fever but thank God no fatal cases. We are overcrowded with the mission on board. We number black and white 53. Our resin of jalap from the Cape is so adulterated it would not act. What I told you of May was for yourself alone. Someone at the Cape raised the report that I was on my last legs and M jumped to the conclusion that he was to stand in my shoes — "He wouldn't play second fiddle to Kirk or Livingstone". I guess Washington was at the bottom of it but say nothing. I sent him away civilly promising to say nothing about a dispute. I navigated from Johanna to Kongone some 700 miles of sea though rather poorly.

David Livingstone

Margins
1. I have written to the Cape requesting whoever is in charge of little steamer to assist any of the missionaries from Kongone upwards.
2. No word from Macrundle Shaw & Co or of the photographic apparatus you sent. We come down again in December or January next. Till then God bless you.
3. My kind regards to all your family with love to your wife — My brother Charles lost £90 by the [failing] of his agent.

---

33. TO JAMES YOUNG

[Murchison Cataract]

Nov. [7/9 ?] 1861

Private & Confidential
My Dear Young,
We have been up Lake Nyassa in a boat which we carried past Murchison's Cataracts at the lowest of which I now write. We left this just 3 months ago and sailed some 250 miles though the Lake proper is but 200. In reaching this today I enquired anxiously if the little steamer had come. It seems as if other slow coaches existed besides the admiralty, but possibly she is detained waiting for a convoy at the Cape. If she is not strong she won't do for Nyassa; if we were to judge from our experience during the Equinoctial gales we should it called the Lake of Storms.

Tremendous seas get up in 20 minutes. We sat for days cowering and looking at the restless sea. Once caught we could neither advance nor recede and anchored in hopes she would ride it out. Really terrific rollers came across with wall like sides to you and having heads but happily none broke over her or we should have been swamped. Our seaman who has been in all seas of the world declared that in subsequent gales no open boat could have lived. Ma Robert would have doubled up in the first storm so whatever Nyassa boat is I hope *she is strong*.

The width is from 20 to 50 or sixty miles. Its shape is like the boot shape of Italy — 200 miles in length at least. No current. Depth unknown accurately. No bottom with our lead line a mile out from shore at 35 fathoms. Then bottom in a bay at 100 fathoms (600 feet) and line broke in coming up though we felt no bottom at 116 fathoms. Plenty of fish and people innumerable. Slave trade brisk. An Arab dhow cut off from us twice to the East coast. We were on the West and could not cross.

People civil, no dues levied nor fines demanded. Here is a field for benevolent enterprise. We are longing for the new vessel and will soon give an account of the Arab one when we get on to the Lake.

We now go down to meet provisions and possibly the Nyassa. Pioneer sadly too long and deep yet carries nothing. Fancy bull's eyes about 8 inches in diameter for cabin and these cannot be opened when she is going. No hold. It was imitation, imitation, imitation, a similar one does well on (Nov 9th 1861) the Thames where people never go below but to drink and no cargo is ever carried. It is good for one river and therefore it must be good for another, but I will not grumble, anything after the Ma Robert — and our Admiralty friend (Capt W) is really anxious to do us good.

We were nearly cut off by a party of Ajawa robbers to whom we went with the hope of being able by a parley to stop the effusion of blood. The Bishop was with us, loaded guns though he did not fire. They surrounded us and poured in poisoned arrows upon us. We had but a few rounds of ammunition and they served only to drive them off. I had not a revolver even and borrowed one in case it came to close quarters. We were trying to pacify the country for the new mission. Caught about 140 captives from Portuguese and gave them to the bishop for his school.

Love to Mama
Affecty D. Livingstone

---

34. TO GEORGE FRERE

H.M.S. Pioneer Near
Murchison's Cataracts

13th November 1861

PRIVATE

My Dear Mr Frere,
Many thanks for your letter of 16th August brought last night by Mr Burrup and the Epitome of news accompanying it. We have heard nothing since we left Johanna and having devoured yours I shall endeavour to give you an Epitome of Trans-Zambesian news. On 15th July last we went up to the Manganja (in mercy refrain from calling them Mangganja try if you can say Mañ-anja) highlands East of this in order to shew the bishop an eligible country for his mission and no sooner did we cross the brow of the plateau than we discovered that the Portuguese had set up an extensive system of slave hunting in the very country to which the mission had come. The first party headed by a well known slave of one of our friends had eighty four captives. While enquiring who gave the adventurers leave to make war and of the captives how they came to be bound, the Tette men escaped into the bush so I handed these and the captives of three other Portuguese parties over to the mission as a beginning of the school. A detached portion of a tribe called Ajawa had been incited by Portuguese who followed the path taken by Dr Kirk from this to

Tette to attack village after village of Manganja, kill the men and sell the women and children to them for calico worth here from 1/- to 2/6 each. We went to try and induce the Ajawa to cease the effusion of blood and came to them in the act of burning three villages. Flushed with victory they felt fully inclined to make minced meat of us all. Some Manganja followers deprived us of the benefit of our English name by calling out that one of their great sorcerers had come and the effect of this in nullifying our declarations of peace was not realised till afterwards showers of poisoned arrows compelled us to act in self-defence. The same system of using the Ajawa as a cat's paw to abstract hundreds of Manganja slaves, and I cannot help believing with the view of thereby of rooting the mission out of the country is still carried on with great vigour. There is an active export of slaves from some one of the ports of the Zambesi probably from one near the Lindi or Massangano. We never knew slaving carried on at such a rate before in the Zambezi. Senhor Cruz the great slave trader has married the Governor's eldest daughter and S. Filipino the younger, so with the French flag to cover it all is comfortable — still suffering from the effects of our Nyassa trip I feel a depression from seeing Portuguese following in my footsteps which neither sun, long marches, hunger, thirst nor even fever ever produced. They have put up a fort at the mouth of the Shire!

We carried a boat past the cataracts in August last and sailed into Lake Nyassa in September. Went along the Western Shore and found it very deep and during the prevalence of the Equinoctial Gales very stormy. 16th November: It is over 200 miles long and from 20 to 50 or 60 miles broad. An Arab dhow has lately been built there to supply Iboe slave export. It fled from us twice.

We got nothing but most contradictory reports about Rovuma. Thanks for your kind attentions to Mrs L. Burrup is a brick he came up the Shire in a canoe the bishop was here on 14th all well.

Where are the Zambesi slaves taken to? Nossibi or New Caledonia. Cruz's export place is a little South of Quillimane say 10 miles the port is Massangano believe me.

Yours Ever
David Livingstone

35. TO JAMES YOUNG

Pioneer in Zambezi

[text badly torn]   19th February 1862

My Dear Young,
As I thought all would turn out right at last so it has come to pass — the MS Gorgon with the brig in tow signalized us on 31st Jany 'I have — steam boat — in

the — brig' then up went the flags again — 'Wife aboard' and my answer 'Accept best thanks'.

We towed in the brig and in a week Captain Wilson and his officers had all the shell and boilers on board Pioneer and two paddle box boats, but we can scarcely steam the river in flood and we are creeping at snailspace and cutting wood of our own.

Rae gave explanations of his conduct which seemed very surprising and bewildering — and [not] accepted by Kirk and me [?] at once. I never sent a word to him in a letter and could not conceive that he had any cause to fear meeting with me had you not given me a hint. He will go on well I think and I believe he has been doing his best heretofore.

As he had the accounts I thought it best to let him forward them to you and so finish up the matter of the coming out. Cutting wood for fuel prevents our cutting any to send by Hetty Ellen.

I did not know that [we] had £400 a year — give a little more to mother, say £50, but hint to sisters that if they can they ought to work if only to baulk that gent who finds some mischief still. And please keep Oswell at Brighton as his chest is tender — this is the case with all children from this country. If necessary to remove Thomas it would be well to send him there too, but he gets on so well at Kendal, if health fail not, keep him there. If Robert goes anywhere I beg you will remember his very exciteable temperament. At St Andrews he complains that he is learning only what he had already acquired at Kendal, if so he would be better off at any of the medical classes. I suppose he has some of my nature and judging from myself placing him in a quiet family for lodging and allowing him to pursue his best in studies according to the advice of Dr Playfair will prove more beneficial than he could be with Mr Hall as a cramming master. Of course I cannot understand or judge of the whole matter — but had Mr Hall threatened Robert's father with a thrashing as he did, a wholesome dread of threatening would have been implanted through his hide for the term of his natural life. Mrs [?] wrote to me about it in great distress and advised me to reprove him for running away. I could not conceive what I should say, so said nothing. It was well he had spirit not to submit to what you never authorized Mr Hall to do, but he ought to have escaped to you or to Dr Buchanan. I believe he went to Limehouse. Wine or spirits will never do for him. Pleasing his companions who are intended for the military profession [may] lead him into a love of them which we may have to deplore.

I think therefore that he must be removed to lodgings and sent to some of the classes required in medicine — say chemistry, anatomy or natural philosophy. He has no fire at Mr Hall's. Let him have control over his own food and fire as we had in days of yore. I think it well to trust to the honour of boys to a certain extent. More will be done when they are aware of being trusted than if they are forced on by a cramming master. As for Mr Hall's [cram] he never once rendered any assistance in the lessons — was once or twice asked but some excuse — 'somebody calling' or 'he was going to take a nap'. There may be a little exaggeration in this but most [assuredly] a person who had been [sent of] one or two charges for Hall's offences was scarcely fitted to secure the respect of any boy. Put your John to Mr

Hall and if he does not pitch into him before three months John has not the pluck I take him to have. Robert feels sore that a letter of apology which Agnes saw before it was sent and says it was a good one, was never answered by Dr Buchanan.

I have never had a copy of the income and expenditure and will be glad if you favour me with one at your convenience. We left in '58 and it is now '62. Let me know also please how much the steamer cost altogether. Captain Wilson of the Gorgon has done everything he could for us. I have applied to the Admiral to have Rovuma explored in boats while we are carrying the 'Lady of the Lake'.

My kind love to Majames. Am glad to hear that Mary Ann is thriving. The height she has attained is no great altitude seeing Majames does not boast of tallness, but take her horizontally and I'll back Majames against all the girls in the house.

David Livingstone

Margin: It may be well to publish a connected account of the Ajawa affair and that I disapprove entirely of the bishop taking the offensive as it is possible his party may put all the blame on me.

36. TO JAMES YOUNG

[February 1862]

. . . Mr Rae has just added that in Edinburgh Robert associates with a bad boy named [J. Cormack]. This is an additional reason why he should be sent out. My pursuits have [pre]vented me giving that parental attention to him I ought, and it might be well to have him some time with me. Please take the foregoing into account in forming [judgement]. If doing well now of course no steps need be taken.

D. Livingstone.

37. TO JAMES YOUNG

Shupanga

5th May 1862

My Dear Young,

With a sore and heavy heart I have to tell of the death of my dear bosom friend of eighteen years — she died on 27th ult and I cannot tell you how greatly I feel the loss. It feels as if heart and strength were taken out of me — my horizon is all dark. I am distressed for the children.

The disease was accompanied by continual vomiting, the most difficult form to treat as the remedies are rejected. You remember how I hastened the first party through the Delta and up to Tette and was blamed for breaking the sabbath to save life. The Portuguese have taken advantage of the sanitary knowledge we acquired and sent their soldiers up to Tette 2 years ago at once. They have lost but two there while formerly by keeping them as Senna and Quillimane they lost nearly all. I prevented too the bishop's party from landing in these same unhealthy months or remaining in the lowlands. All were safe till the bishop returned to the lowlands and perished.

Three months down here instead of three of four days had the Lady Nyassa been able to sail out, was a terrible trial. We have had much sickness, fever and dysentry, but the Lady Nyassa might have foundered had she sailed. Everything was done with the best and kindest intentions and I bow to what has befallen me doubtless under the arrangements of the Almighty Disposer of events.

She was depressed on leaving England on account of Robert's unsettled state and that had an influence on inducing spiritual despondency but happily Mr Stewart, an excellent man, was with her and she derived much benefit from his judicious counsels. It is a pity that Robert was not sent out with her — it would have done both mother and son good. I was talking to Rae last night about it and think that if he (R) persists in refusing to learn, he had better be sent still for a year or so. It would benefit his health but I leave this to you. He may be doing well by the time this reaches. I shall take it kind if you write fully about him. Anything about the children is interesting though unfavourable.

If Robert is sent he would require to be under a kind fellow passenger who would remember his excitable temperament and make allowances. There may however be some new phase of the affair by the time this comes that may render this I have written worthless. Just act according to the best of your judgement.

Rae put the vessel together in a fortnight. I look upon this as quite a feat. I am going down to the sea to bring up all our things there. Dr Kirk & C Livingstone go off this morning to Tette on duty. My love to Majames and all the others.

David Livingstone

P.S. I gave Captain Davies of Hetty Ellen a bill on you £354.7.6. I believe for extra work as arranged by Rae and Captain Wilson.

The binocular can't see a satellite of Jupiter — you didn't try it surely. I can see Jupiter better with the naked eye. None of us can see with both eyes. But looking with one not a satellite appears. Thanks for the lamps. To be tried on Lake.

---

38. TO GEORGE FRERE

Quilimane

19 November 1862

My Dear Sir,

Many thanks for your two letters and accompanying *slips* from the newspapers — the extracts from the Boletim[1] set me to copy part of a despatch[2] I wrote in answer to a request to state for Earl Russell's information what I thought of the correctness of the Viscount de Sá's map. I have not been able to finish it and now send you a line only to say it is under weigh.

We went up Rovuma 156 miles or 114 as the crow flies and touched twice on the slave route from Nyassa to Keelwa [Quiloa]. The Gorgon will do a good deal I believe to stop the stream of 20,000 slaves annually which comes along that road for the Red Sea and Persian Gulph and I am anxious to get up to the source of that trade to do what we can there. We are the better for our sea voyage after passing through that grievous detention in the Zambesian delta — We have lost no time for it as the river is not yet rising — We were obliged to come in here by our fuel being expended when within 30 miles of Luabo and the wind not suffering us to go further in that direction we bore away for Quilimane river to cut wood — We shall tow your mission stores over the bar and they will be taken up overland to the Zambes next month — Canoes cannot go much above this at present — Of course you know that this is not Zambesi though the Viscount's map would make you believe it was — in future please tell the committee always to send their stores to Mosambique, they can then come up to the village which is 12 miles from the bar in Mr Soares' schooner — We could not have got a launch for either love or money but for Colonel Nunes. It is a great risk and there are no insurance companies here so no one likes to risk £300 on Quillimane bar for the sum of £10 or £12. Even Colonel Nunes with all his kindness would have declined sending his launch and very properly so but for the prospect of having it towed out and in — A schooner comes here at stated times from Mosambique and passes the bar at spring tides this is your best conveyance.

Mrs Burrup desired me to get any of the articles belonging to her husband sent home — I had this letter written at the Ruo and gave it to Mr Rae to whom she had applied previously — should you be writing to her please say that I shall attend to her request most carefully.

If the new bishop comes up in July he will find the weather cool and pleasant and will feel so strong that he may do anything. If you can indoctrinate him with the idea that what others have done he may not with impunity do you will render him a service —

Please present my congratulations to Mr Layard when he comes among you and kind regards to your family.

David Livingstone

Private

We have a suspicious looking barque here loading with rice — we suspect she may load with slaves at Makusa a river a little north of this. I have given information to Captain Jago. Her name is Juven Carlotta — We shall capture that slave dhow on Lake Nyassa and condemn it without asking your commissionership's leave. Does not that stir up your bile Eh! It would our friend Dupratt's. Do you know if anything has been done to get Senhor Vianna's share of the price of the Eolus or Eole which our Govt. repaid. We wrote to Mr Dupratt about it per Captain Wilson who was one of the captors?

1. See Appendix III.
2. See Appendix IV.

---

39. TO GEORGE FRERE

Zambesi

2nd December, 1862

PRIVATE

My Dear Sir,

I enclose a note containing the main points of a despatch I wrote to Mr Layard in answer to one requesting my opinion of the Map for Earl Russell's information, I took in the paper of which you were kind enough to send a translation. (I had seen it in the Almanac of Mosambique previously) as it seemed intended to accompany the map and to be to a certain extent explanatory thereof. It is a claim for the whole interior and if I am right in believing that the coast line was allotted to the Portuguese only on the understanding that they were to suppress the slave trade and as they have made a slave "preserve" of it instead, I think there is good reason why the claim should be disallowed. I asked if I might, or rather ought to claim possession of our discoveries in name of our Government and was answered in the negative. I go back to the Shire because there we have gained influence with the natives and should no slaving parties instigate we shall be unmolested. In Rovuma it is beginning at the very bottom of the ladder again. All influence has to be gained and that cannot be effected without time; I am rather down hearted I confess when I see no prospect of being freed from the Portuguese "paper flocade" of the mouths of this river. They never use them while we use both Kongone and Luabo. The expedition ought to be dated only from the arrival of the Pioneer. Give us free navigation of this river and if I have health for one year more the country will be open and a fair cotton trade beginning in an immense cotton field and an annually increasing check given to the great slave market of Eastern Africa. I do not know if any good will be done by publishing the note a good many things I do not care about answering. Be so kind as consult our friend Sir Thomas Maclear, before you do if he thinks it ought to be suppressed by all means let it be. It might only give importance to a map put forth as a claim which otherwise might pass into oblivion. You may not sympathize with my indifference to the publication of what I copied for you with I fear no great zeal.

Mrs Burrup's things are sent. Every attention was paid to her request.

David Livingstone

40. TO JAMES YOUNG

River Zambezi

15th Dec 1862

My Dear Young,

On perusal of your letter of 28th June last I thought it well to write a duplicate of the order to Pay £5,589.5.9 and also for £354.7.6. The original was sent off at once by M.S. Rapid from the Quillimane River where it was dated — and this I put into my despatch box in order that should death catch me ere I come down this river again, you and the trustees may be all safe — Please let me know if the original has arrived. I could send this off now but think that it would go by the same mail as the other. You were quite right to read the extracts you sent me. They are just what I feel now. The only point in which my intentions were more than expected is I expected the cost would be about half, and the voyage out another £1000, or £4000 in all. But it is all right, we have the value of the money to shew, and if we succeed as we hope, never was money better spent.

If Govt won't give a half — say £3000 I would not make a whine about it. I shall hand over whatever of my salary I can save to support the children [and my mother, *added later*] and the interest of the £5000 that may remain will do the rest.

By giving you all I can out of £350 (that is £500 minus £150 for a missionary[1]) we shall get on very well. I think that I shall not live very long and if the children get all that was collected for them in Glasgow and elsewhere they ought to get on.

O if I had looked sharper after my own interests I might have had a higher salary — Burton has £700 (Mr Layard's brother lately got a situation as commissioner at the Cape at £1000 not to be mentioned) and two slave commissioners at Loanda in our Latitutdes get one £1000 and the other £1500.

Bedingfeld had £600 while with me and when I complained a little at this Lord Malmesbury said in his Despatch to me that he thought I had far too little. However I shall [*page missing?*]

I am never to have a home again. All my hopes of doing good in my home among the outcasts of Africa have been dispelled but I shall do my duty notwithstanding.

I wish you to get me a gravestone of granite if it will stand the sun. If not then of iron or stone of at least 5½ or 6 inches in thickness. The shape neat without projecting angles that could be knocked off easily, of good size say 5 or 6 feet high but not with socket above a ton and a half in weight. If you let me know the price I shall send an order to the three Trustees in due form. As also for a watch for Agnes. One sent home I asked Admiral Washington to send to the maker of it for repairs. It was sent out by mistake to me instead of a thermometer Rae lost but I offered to take it and pay the difference. It soon went wrong and hence I sent it home. I can write to you more easily than I could to the others partly because I suppose you will make any allowance for mistakes or for the tone which may seem wrong when nothing is intended and partly because writing to Dr & Mrs Hannan is like confessing to the corner of a confessional box or conversing with a man as deaf as a door nail. They never answer me the idewats.[3]

The water is rising fast and we shall have all the flood for our work. I am glad that you enjoyed coming back to your Lairdship's neuk.[2] Rich old fogies like you and me! Often find they can't enjoy what they toiled and longed for as a competence. Eat and be joyful and drink your wine with a merry heart and send portions to those who are ready to perish and may God's blessing rest on you and yours.

Love to Majames from
David Livingstone

1. See J.S. Moffat.
2. *Scots*: corner.
3. *Scots*: idiots.

41. TO A PORTUGUESE BISHOP

River Zambesi
8 January 1863

My Dear Bishop,

I thank you very much for your kind letter of 31st March 1857 which followed me to this country about two years after it was written. I am sorry that it did not reach me in England for I was anxious to send you a copy of my book but did not know where to send it, and here I have none to send. I mention it now to shew that the kind feelings which you had towards me were reciprocated in my heart.

We left England in 1858 and have been working ever since in the Zambesi and regions adjacent. In 1861 we went up Lake Nyasa in a small boat 225 miles and are now about to carry a steamer past some cataracts on the Shire with a view to launch it on the same Lake, and hope to do somewhat to stop the slave traffic of 20,000 annually who are taken from the Lake region to Quiloa and thence to the Red Sea and Persian Gulph. If the Most High grants us success we shall have done some good in Africa. You can form no idea of the misery that the slave trade causes — for every one taken out of the country two perish by the wars engendered or by the famine that succeeds.

A mission was sent out by a number of learned and pious men in the English Universities Oxford and Cambridge. It was under Bishop Mackenzie who unfortunately died from fever caused by his canoe being upset in the Shire and his medicines and clothes being all lost the disease was not checked — the priest and deacons who accompanied him still continued at their work though they have been much troubled by slave hunters. They are learning the native language and it is partly on their account that I trouble you with this letter. I believe that the Jesuit missionaries translated portions of the Bible or other books into the language of Senna and Tette for I have heard old persons chanting the 'Pater Noster',[1] creed etc. in their own tongue. Now if we could get any of these books in the native language it would be a great assistance to the missionaries for it is exceedingly difficult to learn it as it tumbles out of the mouth of the blacks. If any of your acquaintances knows of the existence of any books such as I mention in the old libraries of Portugal I would very cheerfully pay for copying them and will get them printed and will not forget to give all honour to the old authors — if you can address me — care of Mr Lennox Conyngham — Foreign Office — Downing St London — it will be forwarded.

Tenha pequena luz em lingua Portugueza, e mais gusto para as linguas Caffriaes. Tenho lido a carta da VEa como muito sensacao de coracao — Parece nunca poderemos encontrar na esta mundo, mas temos o meios de communicacao pelo graca da Deus, pelo que nossas desegas subindas voandao como mai velocidade com o corrente electrico, e descem com os bencaes de Noss Salvador em qualque distante lugar. Desejo a VEa a amizade de nossa poderosa Redemptor pedia o favore de lembrancar de mim em sua oracaoes — Ate en contremos. Desculpe este Portugueza [toda].

Minha Esposa tinha tida muito desejo ajudar me na meu trabalho mas depois tres mezes falleceo deixando me sosinho — Reposao os [restos] mortaes della em Shupanga em esperanca da resurreicao de felicidade pela Jesus Christus. tinha sufferida muita pena mas e vontade de Nossa Pae e su reverencia —

Espera que VEa ter melhor [?] para trabalhar por ista Africa. Todas partes tem muita precicao de melhoramento pelo os servidores de Deus.

I am affectionately yours

David Livingstone

As lettras de recommendacao faze muito boa effecta quando chegou em ista Costa — muitos mil gracias.

1. *Latin*: Our Father — the Lord's Prayer.

Translation
Note: *D. L.'s Portuguese is ungrammatical, with many spelling mistakes.*
I have a little light in the Portuguese language, and more taste for the Kaffer languages. I read the letter from Your Excellency with much feeling of the heart — It seems we will never be able to meet in this world, but we have the means of communication, by the grace of God, by which our rising wishes fly with more speed than an electric current, and descend with the blessings of our Saviour in any place whatever. I wish for Your Excellency the friendship of our powerful Redeemer and ask the favour of your remembering me in your prayers — Until we meet. Pardon this [?] Portuguese.

My wife greatly wished to help me in my work, but three months ago she died leaving me alone — Her mortal remains rest at Shupanga awaiting the resurrection of happiness through Jesus Christ. I have endured a lot of suffering but it is the will of Our Father and His submission —

I hope that Your Excellency has better [?] to work for this Africa. All parts greatly need improvement by the servants of God.

The letters of recommendation had great good effect when I arrived on this coast — many thousand thanks.

42. TO RICHARD THORNTON

R. Shire

19th Feby 1863

Dear Sir,

Thanks for trying to get wood cut for us. We are dragging through the part in which you left us.

Excuse me if I repeat that you are on no account to go into a dangerous locality alone. The black men will guard you well only so long as there is no danger.

When we get up we shall require your aid with the steamer and if you go to the cataracts now you might bear in mind to look for a level route for us. We think one will be found a mile or two out from the river.

I am etc.

David Livingstone

(P.S.) I wrote to Clementine requesting him to send our little engine and any parts of the iron house still behind. Please mention in writing that such a request has been made.

43. TO GEORGE THORNTON

Murchison's Ca

22 April, 1863

Dear Sir,

It is with sorrow that I have to convey the sad intelligence that your brothe Richard Thornton died yesterday morning about 10 o'clock. He became very ill on the 11th currt — of dysentry and fever, and no remedy seemed to have muc effect, on the 20th he was seriously ill but took soup several times and drank clare and water with relish. We then hoped that his youth and unimpaired constitution would carry him through — the diarrhoea had nearly ceased — but about si o'clock in the evening his mind began to wander and continued so. His bodil powers continued gradually to sink till the period mentioned when he quietl expired.

He was attended to most assiduously by Dr Meller during the whole of his illnes and as he was aided by the advice of Dr Kirk you may rest assured that everythin was done for his recovery that could be suggested.

Owing to the insidious way in which the delirium crept on him, I regret that h never had an opportunity of expressing any wish respecting his affairs — Dr Meller was by him the whole night. Dr Kirk and I were at hand and sleepless but nothing fell from his lips as last words to survivors.

We buried him today by a large Baobab tree about 500 yards from the first cataract and 300 from the right bank of the Shire — and there he rests in sure and certain hope of glorious Resurrection.

I enclose a lock of his hair: I had his papers sealed up soon after his decease — and will endeavour to transmit them all to you exactly as he left them. The chief part of his property is at the Mission station below this — but it will be preserved with care — mere trade goods sold and everything likely to be valued by his relatives sent home.

He had two men from Zanzibar, Ali and Mabruk, whose wages and expenses home must be paid. I fear the saleable effects will not cover these — He has left some debts at Quillimane — Those to Manoel at Tette incurred in behalf of the Mission and Expedition and amounting to sixty pounds will be paid. It is understood that he ordered goods too for that amount to Manoel — This may arrive in time to countermand that order.

I would have taken the papers, journals, etc. with me at once to Quillimane, but feared that they might get damaged in an open boat. I close this sad letter with a heavy heart and with the prayer that the Almighty Disposer of all events may comfort those to whom its contents will bring much sorrow.

Yours etc.
David Livingstone.

---

44. TO JAMES YOUNG

Murchison's Cataracts

30 April 1863

My Dear Young,

We are working at the road and the steamer is all ready except unscrewing to be carried. We are in a desert produced by one year's slaving and a famine caused partly by drought chiefly by the slaving panic for all along the streams and rivers the "Haughs"[1] can yield crops of maize every three months. The survivors are quite apathetic — cultivating none and actually eating the self-sown stalks of corn which in a month would yield grain. It is a sad sight to see. A few miles from this I passed through over two acres of fine cotton with not a soul to reap it. Rae saw a lot more in the same state. I confess to despondency for these slavehunters durst not go in till we opened the country.

An interim Governor of Tette is said to have been the first to get slaves from the

The Zambezi Expedition

Ajawa by means of some native elephant hunters. The Ajawa hunted each other and one drove the other before it till they came to the Manganja. However it began here we see its effects. I sometimes fear that if these rivers are not to be opened and no effective restrictions put on the Portuguese, I shall yet have to screw our Lady Nyassa away to India and sell her there. I would not ask her price from Government — and only the half in the event of being successful. I wished to make a thanks offering by devoting about £3000 to that which promised to do the most good in the world. This is about double but it will all come out right at last. Dr Kirk and my brother leave by this opportunity for home, each had severe touches of disease, and I let them go willingly, thought I am sorry at parting with them.

Thornton died on the 21st currt — of dysentry and fever. He was knocked up by a trip to Tette and back again, that which nearly killed Kirk and Rae. He went of his own accord letting me know of his intention only after he had left — diarrhoea came on soon after he had returned — then dysentry and he sank of pure fever.

Two of the missionaries died too and if they don't go to the hills all will be cut off. If one does not stir about in the low lands but leads a sedentary life his chances are small. Rae is pretty well and doing well. He writes you of course. £20 were paid you I believe for Dr Kirk. I mentioned it was for a gold watch such as I got from Mr Dent, Strand, for Major Secard a Geneva gold watch [hunter] case I believe and made to wind up by the hand etc. It was less then £20 but if you insure it you may buy the gold spectacles I mentioned or pocket the rest! I just fancy I see how Mama turns up her nose at the idea — but can't she send us a cheese of her own making.

I have asked Messrs Coutts & Co to hand over whatever of the salary is over my expenses for the children. I have given Dr Kirk directions about a Whitworth's rifle. If Mr Gedye handed over £20 [?] which I lent him of private money to Messrs Coutts & Co please use it for the Rifle.

Private

We shall soon be without a doctor for the Pioneer for Meller has evidently got into a great funk by Thornton's death. He would have made a stampede of it. I find that he has been acting the busybody and tale-bearer and making Rae feel queer to us all — but he has found him out. His departure in three months which is his time minus 7 months on sick leave at the Cape will be no loss. Kirk's is and no mistake. Have you a box of birds of my brother which was addressed to you when Rae went home [?]. The British Museum is to have the refusal of all specimens from us.

I hope all your family are well and doing well. We have heard nothing from you for a long time. Sir Roderick has not written to me lately. I fear he may have believed some statements of poor Thornton wishing to be invalided which were quite untrue. Don't spare your pen — you will get too fat if you do.

David Livingstone

1. *Scots*: flat land beside stream.

45. TO JAMES YOUNG

[July — August 1863 ?]

Private & Confidential

I really cannot comprehend Dr Buchanan & Mrs Hannan's refusal to pay after I had sent that legal document written and signed with my own hand. Had you not stood up for me and at a sacrifice warded off the blow both in this case and in that of the Hetty Ellen my credit would have been nowhere. I thank you most sincerely and you must take that document as my promise to pay yourself when I get home. Keep it by you. I felt strongly inclined to write now and break up the Trusteeship so far as they are concerned for they seem to thwart you and I am sure that you did exactly or wanted to do exactly as I wanted in all cases. But I feel it to be a little ungracious of them and to you to act precipitately before I be myself on the spot — the labour being all gratuitous.

But it is rather vexing to see your wish as well as my own set aside with respect to the boys' removal from the south and in regard to the money after you had been at the trouble to get the precious legal "aforesaid" and "other monies" etc. and the upshot "no we will not pay" even for that. I feel very thankful sincerely and heartily so that I leave in God's good Providence such a friend but cannot help feeling dissatisfied that you should be thrown on your own resources in that way, for Thomas Maclear is just such another as yourself but he has not had to sustain my credit as you have.

I trust we may sell the Lady Nyassa well and be able to settle every farthing I owe you as soon as I get home — but I shall do this whether we sell her or not. I feel a little anxious on account of having no idea whether I am to be shelved without salary or not. Lord Russell does not say — but only that I am to come home. If I only knew I could shape my course accordingly and feel at ease. If it is to be no salary it would be better for me to remain out here where I might do some service yet. I should like too to spend some time with my family who have had too little of my care and attention.

Robert came to Natal "pennyless", one box without a key left on board was pillaged on his return to the ship so I suspect he had not been behaving himself. Sir Thomas Maclear a real good friend sent for him to the Cape as he was idling his time away and running into debt in a hotel at Natal — got him lodgings at the Cape at 2/6 per day and I suppose will get him employment at some manual labour. He seems fit for nothing else and must earn some money before any more is spend on him. I wished him to come up if he could do so before December but have told him if he could not that he must get work and stick to it. I must pay some £30 for him now. A hopeful youth certainly, and I cannot free myself from blame in his having so little of fatherly care. I am very heartsore about him.

Should Thomas wish to go to sea in the Navy before I get home be sure please and let him follow his bent. I wish to give them all a good education as my legacy and they may take to any trade they feel inclined for after that. I shall feel it a great kindness if you will remember this without consulting your colleagues. I have not the smallest idea whether Robert will get up or not — it depends entirely on the whim of any man of war captain. From Natal there was not the ghost of a chance of his getting up here.

Rae is well and doing well. The Lady Nyassa is all complete and furnished — we look and long for the rising waters.

I am very much concerned to hear of Thomas' illness. Here we should give quinine for it. Both boys being ill makes one fear some malarious collection under the schoolhouse. I still think that they ought to be in the milder climate of England and in the school at Kendal. I have full confidence but I will not trespass on your kindness to act in any way unless continued illness has been the result. Their uncle Robert Moffat when in England for education stood Newcastle very well but one session at Glasgow college knocked him up and so serious were the symptoms he was at once ordered off to his own country of Kuruman and I have known many like instances. Should the boys however be ill please oblige me once more by removing them south without any reference to anyone. This is your full authority. I hope however they have by God's mercy recovered.

Give my love to Ma-James. I fear that the winter will have set in before I can come home and she will escape the strawberry debt she owes me. Remember me kindly to all the children.

Are you a stickler for the Old Covenant — no prayer for the Queen — no volunteering, no nothing. You ould sarpent, how dare you: Go in I say for union with the Free Kirk if you want to please me and be the model Laird of Limefield — but I suppose it will be only a very small sprinkling of the Reformed Presbyterians that will hold out against the Queen.

The mission is when we last heard on the top of Morambala or 4000 feet high quite among the clouds and no people near to teach — I suppose that the bishop is looking around and maturing his plans, some think to bolt, but I would not say that in public. Between ourselves the High Church missionaries of whom I fervently hoped better things are fit only for well-behaved ladies' boarding schools. They have no idea of what a missionary should do. Still I hope and pray on that they may see their way more clearly. Do you see how gorgeously they go it in Hawaii — South Seas, when the people have already been christianized. Bishop Mackenzie was every inch of him a missionary. All his associates are gone — the good ones dead, the indifferent ones invalided. I would not hurt them by saying a word to their prejudice but they have been a sore disappointment to me and I fear Bishop Tozer's band may be the same. However remembering all our own failings we cannot be harsh in our judgements on others.

I am etc

David Livingstone

46. TO GEORGE FRERE

Murchison's Cataracts

22 Dec 1863

My Dear Mr Frere,

I thank you very sincerely for your kind notes and accompanying slips from the Newspapers which were always very welcome and very interesting. I regret that your last did not reach in time to give you any information for your slave trade report but Cap. Wilson of the MS Gorgon was near and I hope he has given what to him is a new discovery but is an old one to me namely that the slave and ivory trades are interdependent. Get possession of the ivory trade as I proposed to do on the Lake and you render the trade in slaves unprofitable. I tried it though unintentionally in the Makololo country — slave merchants came from Benguella to the subject tribes East of that people and annually carried off large quantities of ivory and slaves — the ivory was purchased for hoes — and Sekeletu having many smiths under him who yield an annual tribute in hoes — I suggested that he should purchase the ivory of the Eastern tribes with them. He did so for the sake of the profit on the ivory and the Benguella traders ceased to go to that district. One of them told me that it was better to get slaves nearer the coast if no ivory was to be obtained for them to carry. The fact of the matter is slaves cost so much for sustenance when a long way from the coast that without ivory they are a losing speculation.

It was a knowledge of the plan I had for stopping the slave trade of Iboe and Mosambique as well as that to Quilloa that made our Portuguese friends frantic. Now slaving goes on faster than ever. Marianno is dead and his people are selling off all the captives whom he caught. It is done openly now and apparently by way of bravado, hundreds of them are taken past Morumbala every week while our mission is stuck up among clouds at its top.

It is about certain that Colonel Rigby's opinion that all the slaves that pass continually through the custom house of Zanzibar are drawn from the Lake district. We came to two Arabs building a dhow to replace one which had been wrecked and transporting slaves as fast as they could by means of two boats. We found that the great slave route for Zanzibar, Quilloa, Iboe and Mosambique goes due West of the Lake towards Katanga — Cazembe etc. We went about 100 miles along it and found that the people there grew nothing else but sustenance for the slave gangs. Fresh instructions were to come from Portugal to render us every assistance. Major Secard when half seas over abused Dr Kirk for the loss of his slaves — and by way of assistance made him pay for a licence to reside in Quillimane — that and making Bishop Tozer pay 4 pence per weight as dues upon calico while mouthing it out loudly for all creation to come and [see].

I grieve much to hear from Bishop Tozer that he intends to bolt out of the country as early in the coming year as possible. The first Protestant mission which in modern times has turned tail without being driven away — With the mission departs the last ray of hope for the wretched trodden down people of this region — and it has the not over creditable peculiarity that while the first party from a conscientious scruple of the late bishop no teaching was attempted the second party never went near the people to be taught, but after drenching themselves a few months on the top of a nearby uninhabited mountain for "strategic purposes" made a masterly retreat. I cannot help feeling very sore at this and see if its tracks are not made for some quarter where all the hard work has already been done.

Cotton on the banks of the Zambezi show a [state of?] which a Portuguese alone can understand. Please present my kind regards to your co-worker Layard and believe me ever yours

David Livingstone

47. TO JAMES YOUNG

Pioneer
Shupanga

[2nd PS]

[10 February 1864]

Water rose on 19th January several feet. We started and carried away the rudder against a bank in coming down. This delayed us for its repair. We are now in the Zambezi at Shupanga 10th February and will be off after wooding and settling some affairs. I hope a stupid tale of my having been murdered has not reached the children. Received a day or two ago yours of August last with papers etc. Sorry to hear of the death of Admiral Washington, a great loss to us.

Robert has shipped into a brig that travels between Cape and New York — as a common sailor. Wm Tod is a brick but I hope we may get enough for Lady N as will pay off everything. You are no less than a brick.

The bishop is off — and we have some of his men on board and about 40 liberated captives — chiefly boys whom he repudiated. Captain Gardner of the M.S. Orestes says he hopes to help us with L Nyassa as far as Mosambique.

I close the letter
David Livingstone

Love to all your family.

PART THREE
Interlude in Britain
1864 • 1866
CHRONOLOGY

1864	July	London
	August	Glasgow, Hamilton with family; tour of Highlands and Islands with Duke of Argyll
	September	Hamilton
	19	British Association meeting at Bath
	23	Speke's funeral
	30	Newstead; starts work on *Narrative of an Expedition to the Zambezi and its Tributaries*. . .
1865	January	Murchison's offer of £500 from Royal Geographical Society for expedition to central African watershed
	March	Livingstone appointed Consul to areas north of Portuguese claims in east Africa
	April	Manuscript of *Narrative* . . . finished, left Newstead; London, seeing book on to press; Royal Academy Dinner
	May	House of Commons Select Committee on Africa (West Coast); at Hamilton, mother dying
	June	Speech at Oxford
	18 June	Death of mother; return to Hamilton
	July	London, Newstead; preparations for departure
	13 August	Left Britain
	September	Arrival at Bombay
1866	January	Zanzibar
	21 March	Arrival on African mainland, Mikandani, north of Rovuma

The Zambezi Expedition over, Livingstone was faced with the immediate problem of disposing of the Lady Nyassa, in which most of his capital was tied up. When he found he could not sell her in Zanzibar, he sailed her at some peril across the Indian Ocean to Bombay. Apart from the crew, he had on board two young freed slaves, Chuma and Wakatini (whom he placed at a mission school), and two men, Susi and Amoda, who had worked for him on the Zambezi: for them he found work in the docks. Having placed the Lady Nyassa on sale, he returned to Britain by way of Aden and the Mediterranean, and arrived in London in July 1864.

British officials gave him a cool reception. The Foreign Secretary, Lord John Russell, advised him to stop his attacks on the Portuguese: Portugal was an ally of Britain, and Queen Victoria was related by marriage to the Portuguese dynasty. In addition to this chill, a cloud of controversy still surrounded the fate of Mackenzie's UMCA mission, with Livingstone implicated in its disastrous failure.

After a week in the capital, Livingstone travelled to Glasgow to see James Young, and then to Hamilton to join his family. He met his youngest child, five-year-old Anna Mary, for the first time: she had been born at Kuruman while he was on the Zambezi. His eldest daughter Agnes was now a woman of eighteen, while Oswell and Tom were at school. Robert, who had tried and failed to join him on the Expedition, had gone to the United States and was a soldier in the Federal Army fighting the Civil War. Livingstone's two sisters were well, but their mother was approaching death.

During this stay in Scotland, Livingstone was taken on a tour by the Duke of Argyll, which included visits to the Highlands, and the Island of Ulva in the Hebrides where his father's family originated. The warmth with which he was welcomed wherever he went in Scotland provided a happy contrast to his reception in London.

On his return from the tour he received an invitation from William Frederick Webb, his game hunter friend of Kalahari days, to spend some time at Newstead Abbey, the Webbs' country residence near Mansfield. With some reservations, Livingstone accepted, and moved there with Agnes at the end of September.

Meanwhile, he attended a meeting of the British Association at Bath in western England, one of his few public appearances during this visit. He addressed a large audience, making a formal attack on Portuguese policy, but the high point of the conference was to have been a debate between Richard Burton and John Hanning Speke on the sources of the Nile. Burton was expected to refute Speke's belief that the river rose in Lake Victoria, which Speke had visited: Burton would argue that it flowed out of the north of Lake Tanganyika. Unfortunately the debate did not take place as Speke died in an accident the day before; but it was clear to all who were interested that the question of the Nile sources was a mystery in need of a solution. Several months later, the President of the Royal Geographical Society, Sir Richard Murchison, wrote to ask Livingstone if he would undertake the task, offering £500 in support of an expedition.

On the question of the Portuguese, Livingstone decided to expand what he had intended as a pamphlet into part of a planned full volume. Years previously, after finishing Missionary Travels *he declared that he would rather walk across Africa*

again than write another book, but no sooner had he settled at Newstead after the
Bath speech than he started work on Narrative of an Expedition to the Zambezi and
its Tributaries. He made use of Charles Livingstone's journals, in fact named him
co-author and gave him the rights to any American edition. (See Appendix V.)
Aided by Agnes and Mrs Webb as secretaries, he finished the 600-page work in five
months, with W.C. Oswell helping him to mould the text into a form that would be
acceptable to the Victorian public.

After leaving Newstead at the end of April, he spent some time in London seeing
the book on to the press, though he intended to return to Bombay and then Africa
without delay. He had accepted the RGS offer, and had negotiated with the
Foreign Office an honorary consulship 'to the Territories of the African Kings and
Chiefs in the Interior of Africa', with which went a once-only £500 grant.
Livingstone would be paid no salary, and could look forward to no pension. A gift of
£1000 from James Young made a new expedition feasible, but at government
insistence, it should not impinge on territory claimed by Portugal.

With his mother dying, he returned to Hamilton, but was not present at the last
breath as he had gone to Oxford to address a meeting, a visit he cut short to attend
the funeral and arrange his family affairs. Oswell and Tom were to continue in
school, Anna Mary would stay with her aunts, Agnes he decided to send to a finishing school in Paris, run by relations of French Protestant missionaries he had
known in south Africa. Robert was a prisoner of war in America.

Throughout this period, controversy continued to surround Livingstone. The
Portuguese had responded to his Bath address by issuing a pamphlet (in English)
questioning his motives and attacking two pillars of his reputation — that he was
the first person to cross Africa from coast to coast, and that he was the first European to visit Lake Malawi, distinctions they claimed for their own citizens. Burton
added his voice to the Portuguese campaign, while at the same time denigrating all
Christian missionary work in Africa. Livingstone replied to Burton in testimony to a
House of Commons committee examining policy towards the mission and trading
colonies which had been established by British enterprise on the west coast of Africa
as working examples of the Christianity, Commerce and Civilisation thesis. The
government continued to support these efforts, but with a strong racialist and pro-
slavery element in official circles, it seemed unlikely that any new initiatives would
be taken to end the slave trade in east Africa.

Only two of the original members of the Zambezi expedition remained in
Livingstone's orbit, his brother, Charles, and John Kirk. He used his influence to
have Charles appointed Consul to the 'Bights of Benin and Biafra' (a post held
earlier by Burton), and after failing to get Kirk to accompany him on his new
venture, worked with eventual success to have him made Consul at Zanzibar.

After a round of farewell visits to sympathisers, Livingstone left Britain in mid
August, after a stay of just over a year. He travelled through France, taking Agnes
with him to place her with his friends, and then sailed from Marseilles for Bombay.
Passing overland across Egypt, he saw the Suez canal being excavated, which when
opened in 1869 would alter the strategic balance in the Indian Ocean.

Livingstone reached Bombay at the end of September, and spent three months *there. He sold the* Lady Nyassa *for £2300, little more than a third of what she had cost, and even this money was lost later when the bank where it was deposited collapsed. He assembled his party, bought stores and animals, addressed business meetings, visited missionary institutions, and made friends with the governor, Sir Bartle Frere, who gave the Livingstone expedition free passage on the ship* Thule, *which the Bombay government was presenting to the Sultan of Zanzibar.*

During six weeks as the Sultan's guest in Zanzibar, Livingstone made his final arrangements and sailed for the African mainland in March 1866.

48. TO —

Tavistock Hotel
Covent Garden
London W C

30 July 1864

Dear Sir,
Will you oblige me by taking into your charge a box which I [have] at Tavistock Hotel and a number of Quarterly Reviews — I intended to have brought the whole myself previous to starting from Euston Square but am too late and hope you will excuse the trouble.

I am &c
David Livingstone

I telegraphed to Mr Young last night that I would go to [Thorburn?] today.

49. TO JAMES YOUNG

Hamilton
2nd Aug 1864

My Dear Young,
Could you meet us at 17 Nicholson's St. on Thursday at 11 o'clock. The children come here on this morning so I have scarcely seen them and wish to spend tomorrow Wednesday with them — its just on post time so I conclude with love.

David Livingstone.

50. TO MRS JAMES YOUNG

Glasgow

29 Aug 1864

My dear Majames,

I have persuaded your worse half to go with me to dine at Mr Ross's this
evening — in order to be able to attend a meeting of Railway [?] tomorrow
morning at 9 & [?] then he may possibly have the affair settled without coming
down again. I am trying to get him to the jeweller's but he won't promise to buy, the
ne'er do well that he is — so be content to sleep alone for one night and oblige

Yours ever

David Livingstone

51. TO CHARLES LIVINGSTONE

Hamilton

2 Sept 1864

My Dear Brother,

I received your letter from Boston a short time ago and having thought some days
over your proposal of joint publication[1] in England and America I feel quite willing
to do anything in my power to benefit you, but I cannot see how it can be done. My
name could not truthfully go on the American work and that I suppose would be
necessary to prevent publishers taking what would be published here and by
republishing it evade your rights in America — It would be the same suppose your
journal were published before mine an English publisher might pirate it but there
would be some remedy at law in this case. I see that Mr Adams has put forth his
book again as soon as I am home with new covers and likeness! I intend going to
London immediately after the meeting of the British Association at Bath and will
consult Mr Murray on the point you have mentioned and will let you know at
once — I am not quite sure but I may go next week. I want to be near London and
have Agnes and Tom with me. I have declined all speechmaking except the meeting
at Bath — and have been obliged to write an immense number of refusals. You
ought to have someone in London to represent your claim in the event of a vacancy
in the consular department. Sir Roderick would do it better than I [*letter cut*] you
get early information and give it to him I have no doubt but he would exert his
influence in your behalf — I shall speak to him about it to prepare his mind
beforehand —

Agnes is here & Oswell too. Tom is over at Limefield as he is better in its high cold situation.

I had a good deal of talk with Lord Palmerston. He and Lady P are the life and soul of all the effort for the abolition of the slave trade. Bishop Tozer has not given satisfaction to his friends in abandoning the mission but in going to Zanzibar he is persuading them that he is only removing it to a better place of operation. Mother is rather poorly [*letter cut to remove signature*]

Margin: With love to all your family. Robert has gone all to the bad & he is in the Federal army. He wrote to Agnes — name 'Rupert Vincent' Co K 3d New Hampshire Regt (8) Army Corps [*cut*] Virginia. If by good conduct he gains the first [*cut*] steps [?] serjeant I shall exert influence to get a commission. This I have told him by letter and you might write to the same effect if convenient. I did not receive your letter.

1. See Appendix V.

52. TO JAMES YOUNG

Newstead

26th Sept 1864

My Dear Young,

Many thanks for the beautiful watch you gave to Agnes, she is very much delighted with it.

We arrived here this evening and found Mr & Mrs Webb all we expected and more — The Abbey seems an immense pile and wonderfully well managed and furnished. I am going up to London to get my journals and begin work after so much play. We were two days at Sir Roderick's at Clifton where we saw Kirk. A man asked me where you were but I did not ask "who are you". This is not much to tell you but the main object of the note is to say "How d'ye do", to thank you for your kindness and bow to our mama — and all the family [?] with that I conclude and am etc.

David Livingstone

Tell Mary Ann not to get up so early as it may spoil her growth.

53. TO W.C. OSWELL

Newstead Abbey,
Mansfield, Notts.

8 Oct. 1864.

My Dear Friend,
I should have written long ago to congratulate you on your marriage but I had sent two letters to the India House & no answer following I conclud [*sic*] that I had not your address. Bishop Tozer gave me your card and from Arthur Vardon I learned that you were Paterfamilias — yes and popular lecturer too at which I mentally took off my cap and bowed my face to the ground. When Col. Cotton I think it was gave me your address while at Bath I had no excuse but I had an overjoy at getting over the dreadful ordeal of a speech to 2500 people — and then by an invitation which followed me to Bombay — and was repeated here, though I declined on the ground that I must do my work near London, and have Agnes and Thomas with me — I came and began to write another book. Mr Webb inviting you to come, I concluded that you might but in the circumstances you couldn't. It was only after receiving your kind letter that I learned that you & the Webbs are personaly unknown to each other, and as I am commissioned to invite you to come at any time — I shall be doing a good service if I can bring you into contact with such charming persons. Good people are scarce so as soon as Mrs O can spare you for a little, come and I feel sure it will lead to your bringing her too some day. I shall certainly run down to be introduced to your better half in a month or six weeks and let you know beforehand. I hope she is getting strong again. Mr Webb has four nice little girls [?] girls like your own and one I never saw till she was five years of age, Anna Mary is with Oswell at Hamilton. Tom is at Kendal but I expect him here on a visit shortly. Agnes is here and woman grown — while Robert I am sorry to say is in the Federal Army in Virginia (if alive) and it gives me a sore heart to think of him. This is an immense establishment — and interesting from its associations with the poet Byron as you know. Webb has his hunting trophies mounted and ornamenting hall and cloisters nicely I am sure a visit from you would give him and his charming wife real pleasure. It would do you good and me no harm. All that [*one line cut away*] beforehand to send to the station for you. I am working with the journal — & both Mr and Mrs Webb are good enough to help in copying and correcting. I hear today that my brother has got the appointment of Consul at Fernando Po vacated by Burton for Santos. On enquiring for Col. Steel in London I found him gone [*one or two words and signature cut away*]

54. TO CHARLES LIVINGSTONE

Newstead Abbey

8 October 1864

My Dear Brother,

I took the sketches over to Captain Need and he is at them with pleasure but nothing of the lip ring appeared. If you have none I shall [ask] Meller to do one for us.

I am getting on pretty well with the work — You have got most of the interesting incidents in your journal and it is a great help — Will you ask Mr Murray who ought to make the map — (Arrowsmith or who ?) I think there is a map in one of the drawers of the Zambesi from Tette upwards. By opening the upper left hand [drawer] you will get a bunch of keys for all the others and one for the wooden despatch box in the same packing case as the flat tin case was taken out of — Open up this despatch box you will get a journal with dark red covers — Which please send at your convenience by rail to Newstead. The box contains also a sketch of the Lake or rather it is in the drawers. The maps need not be sent but be ready for the map maker. If it is to be Arrowsmith you could help him with the names.

Agnes is well and sends her love. Tom is still at Kendal. Waller sleeping ingloriously in a Railway Hotel with his door unlocked was relieved of £7 and his game licence!

With salutions to Mr and Mrs Fitch and family.

I am &c
David Livingstone

55. NOTE TO CHARLES LIVINGSTONE

Newstead Abbey

9 Oct 1864

I open this to congratulate you on receiving the consulship of which Sir Roderick informed me by a note this morning. It is good and may lead to something better as for instance commissionership at Loanda. He proposed my applying for Fernando Po but I proposed you instead and I hope you may be successful in your office — Mrs L will no doubt accede to the plan followed by Burton of his wife living at Teneriffe — If you liked I could give you a note to Mrs Burton and by calling you could get information from her — Some of the statements made by Burton in his books may require refutation & I suppose if you can visit the

mainland you may write papers of interest. I have thanked Sir Roderick & you
had better do so too — I think six months are granted on nomination but don't
know.

---

56. TO CHARLES LIVINGSTONE

Newstead

10 October 1864

My Dear Brother,
In the upper left hand drawer you will find some trinkets of gold which please send
with the other things mentioned in another note which you will receive along with
this.

D. Livingstone

---

57. TO W.C. OSWELL

Newstead Abbey,
Mansfield, Notts.
21 Oct. 1864.

My Dear Oswell,
You must excuse me in troubling you again, but the Webbs would like very much to
have a visit from you, and Mrs W asked me last night to write again and say if Mrs
Oswell cannot come but is well enough to be left for a day or two they will both be
delighted to see you and give you shooting as long as you can stay — I told them
that you liked shooting — and all you have to do in answer is to write & say when it
may [be] convenient to come so as that a carriage may be sent to receive you at the
Newstead Station. Everything however must depend on Mrs O being better, and I
trust she is — Agnes & Tom are here. It is an immense establishment and plenty of
room in the house & the hosts' generous heart. Sir Henry Rawlinson & Sir Roderick
Murchison are to be here next week and they say they would be delighted to have
you too. I am only the "medium".
I got a letter from Mr Moffat — Sekeletu is dead & Mamire a fellow whose knee
you blistered fought with Mamochisane for the regency and Mamire fled to

Lechulathebe & came back with the Lake Ngami folks to steal the Makololo cattle and so things go on. Our Robert is in the Federal Army. The American Minister was here to lunch & promised to see to his liberation — he was kidnapped he says — With kind regards to Mrs O and family.

David Livingstone

58. TO JAMES YOUNG

Newstead

30 November 1864

My Dear Young,

Please to thank the young man who did the fiddle for me. You were the last who wrote you say if not you certainly have the last word.

I have been thinking over what you propose to ask — Sir Roderick thinks that they won't give the help or consul's salary without consulate. In my mind I shall get nothing but a refusal and that I don't like to run the risk of getting.

Lord Pam[1] when spoken to in a private way says it is Lord Russell who should be applied to so I shall go out again — spend £800 at the Cape and die leaving the rest for my children.

Sir Roderick says I might finish up by settling the watershed of the Nile question — a purely geographical question has no interest for me.

I want to do good by opening up a path inland North of the Portuguese. My time is passing away very quick. Some of the book is in the hands of Murray for printer.

Kirk is invited down here. He is looking out for some situation himself and I have been trying to find something for him too in a quiet way.

The folks here have saved me a good penny — Agnes gets lessons in music from a first rate performer — governess to two little girls.

Tom is not well again and Kirk wants him up South. He must get education for that is to be his wealth but his complaint is a drawback as men talk but it is all part of a Father's discipline and of course the best.

I have eschewed all feasts and gulravagings[1] in London though invited to many parts.

Saw Miss Coutts whom I told you had sent out a sort of preparatory mission to Borneo.

I wrote to Adams the American Minister today asking if he had done anything for Robert. I think he has neglected it.

How are you all? My love to Majames and the new edition or as the psalms say "another of the same" called Mary Ann.

My brother's wife it seems has come.
David Livingstone.

1. Lord Palmerston.
2. *Scots*: merry-making, carousing.

59. TO JAMES YOUNG

Newstead
4 Jany 1865

The same to you my Hearty. I felt inclined to write some time ago and call you an old Heathen Turk or Infidel or something else you deserved for being mum so long but thinking that perhaps it might find you in the dumps and set you "acussing and swearing dreadful" against Majames I let you alone.

Glad to hear you have got rid of B.

Did you see a speech by Mr Gladstone in the Times, I think on the day before yesterday — at Wolverhampton — in which he puts forth the "Gas idea". How is that getting on? I have been doubting of late whether you can drive gas down to Glasgow without having steel pipes — see how they are always bursting in London and if high pressure is put on won't they be the more liable to burst. If you were lower than Glasgow I should say all right.

The patent would not hold water as I am not the first who applied the stuff — so a lawyer tells me. He says the only way would be to get a large lot of the material and then sell it at a large price. The nut contains that preservative quality in the shells alone — The kernel is eaten but contains so much oil that it is not wholesome to eat much. The shells at the present are of the same value as those of the hazel nut and all thrown away — so one would require to import the entire nut, get them shelled in one part and the nuts used for its oil there — the shells sent to another part and there boiled for the preservative liquid or resin — and I having none of the commercial quality or faculty could not manage it. My *specialité* is among black folks.

The walls fell down because the rest of the bricks were not set as well as the first. None of my houses fell down and I built five. Besides what could you expect? If you had given us a cask or two of champagne not a brick would have budged for ages. It is entirely your own fault. The stick is your own as I mean to give it to your better half and advise her to use it instead of clawring your head with the usual cutty stool.[1]

Agnes has been dancing like mad, at two balls. A colonel's wife here took her to the first at a neighbouring laird's, and that laird's wife took her to the second!!

Tom is still a little troubled with his complaint. I think I had better take him out with me and place him at the Tea plantings in India.

The adjutant of Robert's Regt. says he is sure that Robert was not left dead or wounded on 7th Oct, the day he was captured on a skirmish before Richmond. They are negotiating for exchange and will give him the "consideration his family merits".

The Adjutant thinks that I spoke disparagingly of their cause at Bath. I didn't.

The Portuguese are mad at that same speech and I humbly hope they will be madder and red hot lunatics when they get my book. I held forth last night at Mansfield and am today empty of all the wind.

David Livingstone.

Margins
1. I am glad you have set James into the partnership. You have John and Graham next to provide for and had better keep places for them instead of having any strangers. If they turn out fools it's because they inherit it. At least that is what I think of my own case. And there is no saying what Majames may do yet. If you speak her fair I would back her to agree to anything you like. Twins even.
2. Kirk is here [awhile ?] and did very well at the meeting last night. My kind love to Mary Ann and her mother.

1. *Scots*: scratching your head with the usual short straw.

---

60. TO JAMES YOUNG

Newstead

7 Jany 1865

My Dear Young,

You will have seen that I held forth at Mansfield. The papers report very meagrely what Mr Webb said and I am sorry for it. It was to the same effect as Sir Roderick at Bath — I have promised to go out as soon as I can get through with this work but not purely geographical as they want me. I must be a missionary. It was a great affliction to be a "skipper" but I had to be or let B[edingfeld] ride roughshod over us all.

I see nothing for it now but sell Lady Nyassa for the monsoon will be changed by March and blow from Africa to India. I must spend that £800 I have at the Cape in my next trip and it must be with Indian Baloochees I go. Take Robert with me if he gets off, but I need not bore you with my affairs. You asked when I could come

North. I am at the proofs, that is all I can say. Sandy Bryson's watch glass is breaking, it would [be] like the Yankee who borrowed a wheelbarrow and came back with a request to have it mended as "Father wants to borrow it again next week" so you had better buy it for me *cheap*.

Let us see what you deliver yourself [of] when you do come out *strong* on gas — can you pump it down? Lots of [?] coal in this quarter. Gas is made for the house. We have been trying trout spawning in our leizure moments, 130,000 ova down.

Love to Majames etc.
David Livingstone

The gas I emitted in Bath sounded in America. Ugh!

---

61. TO W.C. OSWELL

Newstead
Saturday Eve
8 January 1865

My Dear Friend,

Mr Webb on being told that Mrs O was so weakly, said this evening "Will you tell him to bring her here. We make all who come well". I am sure he really means what he says.

Mrs W is better — able to get up — and if you could bring Mrs O for a fortnight or more you would only gratify them and benefit your wife. I would not urge you if I did not see clearly it would be a benefit and agreeable to both families. It is an immense house and you could do as you like in your own apartments or elsewhere, and it is really a healthy dry spot, though there is lots of water about, and we are spawning the trout like mad. We have 130,000 ova down.

Now for your wife's sake come and I'll give you advice worth any three!! doctors in London. It is change of air to a mild spot like this she needs. I who am a good doctor spoiled must at times blow my own trumpet.

Thanks for kindly promising to work a little. The sheets will come by and by. I set Agnes to write being busy myself at the time.

David Livingstone

62. TO W.C. OSWELL

Newstead

8 January 1865

My Dear Oswell,
I wrote you this morning and Mr W sitting in the parlour today "Now do tell
Mrs Oswell that I am an invalid too but hope to be well presently, and she can do
just as she likes, lie in bed asleep as she likes — come to meals or stay away — or
occupy this drawing room all day". We both assert that this is a capital place for
invalids.

Webb said "Do you think Oswell". I can't tell of course, but think if you can take
her to London you might make a trip here and then see what the Muffs[1] in London
say on the way back. I am quite sure that you will never regret that I urged you to
bring Mrs O here [quarter page cut away] rest are charming people.

He is very fond of telling about Africa and everything African pleased.

Now I think I have done my utmost and so place the case in your hands. You
possibly may have reasons which I don't [quarter page cut]

1. Idiots.

63. TO W.C. OSWELL

Newstead Abbey
Mansfield, Notts

11 Jany 1865

My Dear Oswell
The contents of your letter to Agnes has been duly communicated to her venerable
progenitor, and you are *in for it* for I sent off this moment a proof of the first
chapter to Prof. Owen with the request that he will send it on to you — This is
pretty well poked I think — but I shall send to-morrow proofs before I go over
them — and anything very glaringly wrong you need not mind for the printer
points to them as unintelligible and very often it is his own fault. All these I
shall correct when they come back from you, but the minuter errors which you will
see in this in a smaller hand between the lines are what bother me most & any additions needed to make it more clear, point out; sorry for the cause of your being

unable to come as it would have done Mrs Oswell good — but it is I see out of the question at present —

With kind regards

I am &c
David Livingstone

I wrote to & received an answer from our friend Mrs Vardon.

---

64. TO W.C. OSWELL

Newstead

12 January 1865

My Dear Oswell,
The proof came from Prof. Owen. I shall send you one of the first impression on Monday which will enlighten you on the way to correct. It is for [ordinary] purposes quite simple. [?] on the side draws attention to everything. ✔/ is Greek δ with the upright line and means *dele* or take out. If you would prefer after all to let a word remain in you put dots under it thus or put stet or simply st on the side and the printer is expected to follow the point [of] your pen wherever it goes and insert accordingly. ,/;/:/. Shew the points) or / draws attention to anything you may put between the lines. A line <u>under</u> italics, R/ Roman and if you want to transpose you do so and put ⎵⏋ on anything to draw attention.

Thanks for what you have done and hoping [*one-third page cut away*] and I hear gave great pleasure.

Hayward one of the Saturday *Revilers* left us this morning. Wasn't I civil — that's all.

A pond is running off and we spawn the fish so I am in haste, but with kind. [*one-third page cut away*]

65. TO W.C. OSWELL

Newstead

21 Jany 1865

My Dear Oswell,

Today I send pages 69—80 and 81—92 for your correction. If you can let me have them back again by Tuesday or Wednesday I shall be glad as I wish to send them up as soon as possible. I have gone over them once — Have not yet gone over your last which came this morning.

The contents are very good but I want please a great deal more of them. They are to be at the top of the chapters. For want of them we have gone on an everlasting Chap I. I have not gone over your last with your corrections yet. I don't like long nebbed[1] words though from our Scotch I may put in too many of Latin origin.

You had to *investigate*, a capital word. Prof. Owen had to *note*, a capitaller.

I send you also the same sheet you sent me today but uncorrected that you may make contents for and divide chapter. I think the departure from Tette up to the Makololo country might be one, Return another etc. The sheets in front of page 57 are only for contents. You will get them by and by paged [?] is corrected and paged from my slips. Anything that the natives of this country won't understand (the stoopids) you will do me an immense good by throwing light on etc.

N.B. pages 29—56 are sent simply for contents. What I made were up to "Manganja country well watered" which may be in the part torn off. If you wade through this you will think less of the writing. Some of the sentences puzzle myself.

But all 29—56 have been corrected so far and only want your chaptering and chapter contents. A sentence about [?] Paragraph will do. If you think a paragraph ought to be made write PP/

Your corrections for 29—56 will be expected when they come back to me in another form than those slips. With kind regards to Mrs O. I am etc.

David Livingstone.

1. *Scots*: penned.

66. TO JOHN MURRAY

Newstead Abbey

24 January 1865

My Dear Mr Murray,

By this day's post I saw some first proof and a part of the 'first revise' and have to trouble you for pages 99—100 of 'first proof' which by mistake has been left out and 7—8 put in duplicate instead. If I knew Mr Clowes' address I could send to him direct — I have inserted some of the drawings [for] woodcuts but don't know if they can be separated as for instance — the hippopotamus spears and Manganja basket — dress and spears. I shall point to their insertion elsewhere & if they can't be disjointed no harm will be done — Will you please let Arrowsmith have a proof for his use — I shall have a lot of proof to send by Friday next and am &c

David Livingstone

have not an idea of what will do for Frontispiece. Would a bird's eye view of the Falls do?

67. TO W.C. OSWELL

Newstead

25 Jany 1865

My Dear O,

es, because up to 56 the first proof has gone to the printer corrected so far as I could and will return for your polish paged as "First revise". "Manganja well watered" is where you have to begin with contents for the head of each chapter and chaptering say Chap III for I shall point out Chap II at "Arrival at Tette" Reception by the Makalolo" will be beginning of Chap II.

First ascent of River Shire may begin Chap III. Second ascent for Chap IV etc. and put some of the contents of Chap I forward to Chap II.

The journey to the Makololo country may be divided into ascent to falls, sojourn at Sesheke, Return journey to Sinamane, Return to Tette, to Kongone, Pioneer etc. and up Rovuma, up Shire again with bishop Mackenzie.

I this day send you 129—148 to keep you going. It is all that I have printed as yet and woe is me if I have still to write a lot. You had better not hurry too much but do as I do in travelling not make a toil of what may be a pleasure (if wading through nonsense ever can). Oh! I didn't know at first that the ox had died of disease. I found it out after we had eaten it and then like a wise man hid it from you so you need not be so angry after all.

D. L

25 Jan. 186

Fragment film with 67
p.1 We hear that our old friend Col. Steele is to be married to a pretty American a Pau — France.

Coming back to the ship. You did not notice perhaps that the contents are alread down at the head of the chapters i.e. Introduction [?] Chapter I. After case c voluntary slavery and visit to Kebrabasa cats a new chapter and . . .

p.2 to contents. Mine are up to "Manganja country well watered". But whateve yours may be I can work them in. Would "et Timeo Danaos et dona ferentes"[1] do i the text at the end of the "other" story. I can alter striped pig to say another thin quite as true namely reverting to original wild pig of India. What is its prope name — *Borus Indicus*[2] or *Pigus abominationibus*[3]?

1. *Latin*: And I fear the Greeks even bearing gifts. (The first 'et' is not Virgil's.).
2. *Pseudo-Latin*: Indian boar (or bore).
3. *Pseudo-Latin*: Pig abominable.

68. TO W.C. OSWELL

Newstead Abbey

3rd Feb 1865

My Dear O,

Today I got a lot of corrected matter First Revise from Mr Murray, and Mr M knowing that you were doing what you could proposed to stop further proceedings. To this I replied go on but look at our 4th Revise perfect more perfect and most perfect but the multiplication of commas — and some little things do look better. I told him he had better go on with his friend and when he had corrected I would incorporate his with yours and then send on to the printer myself so as to avoid confusing him.

Please send Mr Murray the 4th Revise of the beginning which I sent you if it has not gone to the printers. If it has will you say in a line to Mr Murray that it has gone on.

I am etc.

Your brother confused and nearly worse confounded by not getting back my Kebrabasa stuff.

D.L.

He suggested telling a little about steamer Pearl and MaRobert which was good.

69. TO MRS CHARLES LIVINGSTONE

Newstead Abbey

3 February 1865

My Dear Sister,

I saw Mr Murray on Wednesday and found that he had received a letter from you saying that you had failed to negotiate with Trubner & Co, Boston.

He says that Samson Lowe & Co. the agents for Harpers are very anxious to get the book though Harpers pretend to you they are not. He has also written to Harpers telling them that they must make sharp and arrange with you or they will lose it. Mr M will send the stereotype plates and pictures so that no time will be lost — You have not yet received any because they are not ready. If Harpers won't do others will. I think I may have the first sheets all ready with 2d revision in a week and Mr M will not delay but send the plates to you at Boston I suppose — I have been rather late in getting this written in fact forgot it till the post is gone and Mrs Webb is to send a servant to catch the mail cart.

Agnes is well & sends her love to you all.

I am &c
David Livingstone

---

70. TO W.C. OSWELL

Newstead Abbey,
Mansfield, Notts
8 Feby 1865

My Dear O

All right — It wouldn't be natural if too well done — and as for the critics I shall be away & not hear what they say — If plain and readable that is all people can expect from me — I thought I had sent pages 25—36 but yesterday got 37—48 first revise back from you so send 25—36 again. If it needs no serious additions from me — send it after you have done your best to Messrs Clowes & Sons, Duke Street Stamford St, Blackfriars, London.

If any addition is required send it to me I send him today 37—48 with your corrections.

I spoke to Waller one of the Mission O & C. He does not like my telling that disapproved of Bishop Mackenzie fighting — although he admits that he was present when I tendered my advice against it. The reason why I am so minute is another called Rowley wrote a letter to Glover at the Cape putting all the blame on me & this was printed in the Times & all the Papers and though [Rowley &] all of them disapproved privately of what Rowley said yet nothing came from them privately. I send some more First Revise to-day and am &c

David Livingston

71. TO W.C. OSWELL

Newstead

9th Feby 1865

My Dear O.

I send by this post printed contents the last part of which I altered a little — and you will please alter if you think anything necessary.

I send also the first of the third Revise just to shew and on its way to Clowes.

I am expecting back pages 99—100 at your convenience and pages 25—36 I think, which I sent yesterday either to [?] or to the printer. I am not idle though I fear that I may seem pushing you too hard.

Clowes address is Wm Clowes & Sons Duke St., Stamford St, Blackfriars L.

D. Livingstone

Don't fear your doing any harm. I don't wish it to be anything but plain and clear — and that you help me much [by] saying where it is muddy.

72. TO W.C. OSWELL

Newstead Abbey

13 February 1865

My Dear O,

Thank Mrs O kindly for writing the note yesterday — I changed all the three you mentioned — soldiers — east coast — courageous [companions] — but held out like a brick against the semi-colons being tampered with — I did not like to say all my companions had courage and perseverance because some had neither the one nor the other but I put it all on your conscience.

Perhaps "torpidity of skin" came as an afterthought in the proof after it came back from you.

I alter some little points in the contents for instance "Travelled monkeys" is very good, but in me it smacks as if I were sneering at the poor fellows.

All went off to Clowes this afternoon namely 3d Revise and 25—36 with your contents — this past post time and I am tired.

18th. I thought to send this off three days ago but have been so absorbed in manuscript and correcting that I have got quite muddled and dazed. I send paged matter up to 48. I think if it succeeds it will owe more to you than to me.

Some passages in the first proofs I don't understand myself so I blame the printer.

You will see specimens of the illustrations [?] it's the worst as I can't get men to make blacks like their photographs — take high cheekbones, low foreheads — big lips — mouth from ear to ear — ears like an achter[1] wheel, mix — [?] — that is the prescription of all artists except Angas.

Agnes is delighted with your photo — you look better than ever thanks to your better half — I am very old and grey — and face wrinkled like a grid iron. A barber offered to dye my hair for 10/6 — I must be very good tempered for I did not fight him.

I found Tozer not in high repute in high quarters. It is known how he would have left some 30 or 40 boys and five widows whom I brought away — I must mention it but will do it as gently as I can. He wrote to Bishop Grey (a prince of a man) that I took those poor things out at Kongone with closed hatches! MAKAHELA.[2]

I left out all the passages you noted in the proofs. Waller is writing a paper in MacMillan which you had better look at — It is good — very — It was in proof only I saw it. I advised him and think of getting him to write one for some other mag.

What is the game we used to play by keeping a ball up by the hands or by [striking] it against the end of a house — Fives or golf or what I am forgetting my English? They throw up the ball by striking it against the ground and as it rebounds catch it.

With love I am etc. DL

Saturday morning
I send 17—48 about perfect I think so far as the power of the author extends.

1. *Dutch*: rear wheel of wagon.
2. *Tswana*: nonsense.

---

73. TO W.C. OSWELL

[18?] Feb '65

Notes
Change introduced about old English slavery puzzles me but printer will know.

All the underlined passages are deled the side lined passage [[==]] [?]

I must say a little to butter the Americans as I think the blackguards will do something [?] themselves yet take Cuba perhaps I have put a question to that effect. Milton is not worth much I think. He but rode back to the starting post again. Often it is stated that neither saddle bit nor bridle are used. I have answered about Boma. Chapter is reduced — all up to 233 are off today 18th to be paged.

David Livingstone

[If I] don't see them after you do — the mile cut is in — you have not seen it.

74. TO W.C. OSWELL

24th February 1865

Dear O,

Only a line to say 83—93 received with thanks. The Old D. is simply if you found [?] need by reference to the past pages — but you need many more for that. To be burned and for nothing else Mr Murray has got a "literary friend" to make corrections on the sides. Who this may be I know not, but he puts in many more commas than we did and makes some things clearer — so I say nothing. He asked me if they would do so of course I said they might.

I shall need to look over dates in the final Revise. I am quite muddled now sending proofs corrected and then lighting on another and correcting again. I am going insane I think with this writing — for the Yankee papers have it that I married a lady at Constantinople which place I never saw.

D.L.

75. TO W.C. OSWELL

1st March [1865]

Today I send 149—164 First Revise — of which the principal alteration is in turning Portuguese stuff into a footnote. Then 165—180 also 1st Revise in both of which if you please to make what alterations you can and send on to Clowes. I send one of 165—180 at same time on to Prof. Owen and he will send it to you and you may be good enough to incorporate anything he may put down in yours.

Also 3 Revise 17a 24a and 23a—32 and 4th Revise of beginning. Surely it would do now. If so please send it to Clowes.

Contents of Chap II have not been sent to printer I think. I intended to leave for London but our friends here won't hear of it.

25,000 trout spawn down here — this is more congenial than correcting but the one is pleasure the other duty

No signature

76. TO W.C. OSWELL

Newstead

8 March 1865

My Dear O,

I return you 83—89 2d Revise that you may just glance over what I have inserted & see if it pleases you.

The Literary friend I suspect has become grumpy, for what I sent Murray has returned without a word! It does not matter in the least. I send you a part of what he did that you may see he is not so very far above us as we might think [?]

The printer charges every word so if he is angry he is foolish too. I go up to town on Sat. and will arrange to give him all the final Revise a good way [on] on Tuesday next so if you can let me have it back by that time I shall be obliged — I go to see Mr Layard [?] he wants to tell me something on my affairs.

I am &c
David Livingstone

1—12 3d Revise is also sent in duplicate — please keep a set for reference — the old ones if you first glance over to see that corrections are inserted, need not be returned. I send the account of Kebrabasa to Prof. Owen today. When it comes to you will you put in what corrections from him you like and leave out the rest then send the [?] copy to me for Clowes — I am at manuscript tooth and nail finishing up that is the reason I put so much on your broad shoulders.

I may be wrong about literary friend — it may be [?]

77. TO W.C. OSWELL

Newstead Abbey,
Mansfield, Notts.

21 March 1865

Dear O,

Yes — A Captain Fraser is going in for slave labour on a large scale — he to receive $\frac{1}{2}$ profits with the Sultan other $\frac{1}{2}$ — the latter to do all the "licking" I put out Zanzibar.

You need not pretend to stupidity I bear the palma qui meruit.[1] I wrote you a long letter & never saw it again — then after failing to find it yesterday wrote a few lines — rushed off to the letter box in the corridor "just in time". When just time to dress for dinner I found a mare's nest — often observed that slight but long continued pressure by the lip ring made the upper gum & teeth go in wards — so we have a cure by a suitable appliance for ladies with too prominent teeth — put that in the Revise as a query for medical practitioners & waked from my dream by finding the bell is just going to ring and there is the letter before me I had posted in such a hurry 3 hours before.

I see many things not put in by printer but have set "Literary friend" to watch for the future.

David Livingstone

Margins:
I have put in footnote at Falls
Think you are wrong about [?] but refer to Kirk

1. *Latin*: Let him who deserves the palm bear it. (In full 'Palmam qui meruit, ferat.'): Nelson's motto.

---

78. TO W.C. OSWELL

Newsted Abbey
Mansfield, Notts.

23rd March 1865

My dear Oswell,

In some unaccountable way I did not notice that you thought you would be able to come down for a couple of days — I put the notes down with a book intending to go over them all carefully when I got an answer from Kirk about the Falls — that seemed to be the most puzzling question — and somehow missed in a sort of dumb wonderment to observe all you said — but it does not much matter only you will be savage at me for not saying a word.

Mr Webb got his knee locked i.e. a little bit of cartilage slips into the joint, six days ago and has of course been laid up ever since — then Mrs Webb has got a bad cold, and again he yesterday got a bad sore throat and cold. A Col. Oakes was coming and he was telegraphed to stop him — so it is likely that they would have begged you to put off for a fortnight. I shall speak to them tonight but you must place all the blame on me.

The fact is I set to on Monday and in desperation sent off in Indian file one revise after another — contents bechaptered and everything else.

I don't altogether approve of the literary friend for instance — "Should sleep

neither hungry nor thirsty" he makes "should neither hunger nor thirst". I left i
though I think mine better —

It is the printer that makes the marks as untameable. Hippo may be said to res
not sleep at the bottom.

Tom's complaint I have today ascertained from Dr Watson to be blood an
albumen from the kidneys. The latter not more than the quantity of blood wi
account for — My friend took Oswell and him from school in England against my
will and put them to a miserably ill drained school near Glasgow. The consequenc
was they were both home in three months ill — O with inability to read except by
holding the book at arm's length — He got over it in a few months but Tom's urin
was as dark as Port wine for a long time — Has had it for some two years. It is les
now but I sent him to London a fortnight ago and Dr Kirk, (a good fellow) wen
with me to Dr Watson.

He is not to be in cold or damp and Dr W thinks a warm climate will do hin
good. I think of taking him to India and possibly settling him to tea planting. He i
a steady good fellow — Robert rebelled against the same Trustees — and the en
is where he has found himself (if alive). I do not blame him very much. No mor
word of him. The Yankee Adjutant of his reg said to Mr Adams the America
Minister, that in my speech at Bath I had not spoken well of their nation — What
said was that these terrible wars always taught terrible lessons and perhaps th
lesson this one would teach would be that "though on the side of the [oppressors
there is power, there be higher than they". It is a line from Ecclesiastes — I did no
answer this fault finding for fear I should make bad worse. I have heard nothin
except that they are exchanging prisoners — I should like to take Rob wit
me — Then again that same speech set the Tagus on fire and the Portuguese got
fellow[1] to write me down in the official journal Diario de Lisboa which if you don
read Portuguese means "The Diarrhoea of Lisbon". They quote Rowley to prov
that all disasters arose [*quarter page cut away*]. I think of leaving next week fo
London — I think that is the only way of getting out of the confusion with th
printers — Be at them and poke them up with a long pole. Contents not come ye
nor has the part about the zebra. [*quarter page cut away*]

1. See Lacerda.

---

79. TO A. SEDGWICK

Newstead Abbe

26 March 1865

My Dear Friend,
I thank you so very much for the long, kind and sympathizing letter with which yo

favoured me three days ago. It quite warmed my heart and I assure you that I feel most grateful to find that amidst your bodily trials and weaknesses you laboured to afford me a fresh proof of your friendship. I heard much about you from our friend Mrs [Ogulthorne?] at Bristol, and would have written you long ago but all my time is absorbed in writing another book — I am literally over head and ears with copy, proof, revise and copy, proof, revise again till my eyes are sore and my brain all muddled. I have imposed much more labour on myself than I might have done, by working in my brother's journals into mine, so that his family in America might obtain any profits that may be realized there — And all the while I remember that I have my vessel waiting for me at Bombay, and my men wondering what has become of me. I should have gone off and left it unfinished but for the belief that in this work I am affording information which may lead to the Governments of the [world] putting a stop to the farce played so long by that of the Portuguese on the East Coast, of pretending to dominion while doing nothing but shutting it up from the rest of the world as a gigantic slave Preserve. The hope that I may do a little service in this way, alone has kept me here.

I fully reciprocate your sentiments about the unhappy appointment of good bishop Mackenzie's successor. It was a sad mistake to send a man who was not prepared to make a stand against the monstrous iniquity of the country. Our late friend had made the first stage to success — He had gained the confidence of the people as a true friend to them — and an enemy to slavery. The Portuguese were ashamed of their deeds — they would never have said a word to Mackenzie about that liberation of captives which won all the people's hearts — they never grumbled even at me. And this first step which in many missions took years to accomplish was by good providence gained by him in the first twelve months.

"By Sebituane" I overheard one of the Makololo say "Had the bishop lived we should by this time all have been in the book for he had a heart."

You may if you can imagine our chagrin when the successor came, praying for success to the Portuguese in all their undertakings and in reference to the poor children whom good Mackenzie would have trained up as members of a Christian family, say "I *repudiate* the acts of bishop Mackenzie, one black face is as good as another one." That certainly was not the spirit in which Cambridge sent forth her mission. I cannot refer to the natural consequences without a sinking of heart. The mission placed on the top of a mountain where no people lived to be taught — Then bolting out of the country to a place quite as unhealthy as any part of the Interior, where comforts may be enjoyed. Fancy Mackenzie or let us say St Augustine bolting to one of the Channel Islands by way of Christianizing the pagans of the interior. It was a sore disappointment to me, and I fear a precious opportunity of planting Christ's gospel lost through sheer want of pluck.

I would take a long time to tell you all I feel. I am living with a friend whom I made in Africa, Mr Webb of Newstead Abbey and cannot move till the book is done. I am pretty much through with it and then after a short visit to my mother in Scotland am off again to try and get a way into the Interior North of the Portuguese. My appointment gives me wide enough scope and if I saw a beginning made with lawful commerce and Christian missions I think I could lie down in my grave and rest in peace. I shall send you the book as soon as it is out. I congratulate you very sincerely on attaining the goodly age I shall never reach and pray that the Almighty may keep his everlasting arms underneath and around you. May his blessing be upon you. And I thank you kindly for sending me yours. If you are writing to Canon and Mrs Guthorne remember me kindly please to them. I am about to say that there appears to have been no "stone age" in Africa. Always one of iron and following Archbishop Whately I infer that the Africans had a super-human instructor, but I will write again. This is only a note on Sunday Evening and with much love I am

Affectionately yours
David Livingstone

80. TO W.C. OSWELL

Newstead
1st April 1865

My Dear Oswell,
It seems it is wrong to say Mrs O or Dr L so I begin to reform on this Allfools day. I made a dreadful [?] — made new contents for the journey from Tette upriver and found after I had been at the trouble I sent them off several days [4 words] beside me in print I looked only at the upper chapter in the slip.

I send paged matter up to 144 that you may see the form they assume. The last three sheets I have not corrected except in the copies I have already sent off to Printer. If you see anything really glaring I could stop the press but I think there can't be much the matter except in the three sheets I send. I send them with pictures that you may mark on the pictures between what pages they are to go and see them at same time I had to ask Mr M [one-third page cut away] about our stay in Sesheke and visit to Linyati. I ought to have sent it sooner and will be obliged if when you get it you hasten it to me. There is no hurry with the paged matter. I have left in chasm as you suggested and alter channel [one-third page cut away].

P.S. How would I do to put all about the Oxford & Cambridge Mission. The part of Mrs Oswell's quotation had gone to be paged but I will put it in it is so appropriate.

P.S. [?] one chapter and head it so leaving out many of the dates.

P.S. [mutilated]

81. TO W.C. OSWELL

Newstead

3 April 1865

My Dear Oswell,

I put about Major Secard in an appendix to have the narrative of the Lake continuous and [?] controversial matter, and am now doubtful if it is not shelving what ought to be seen prominently. I sent paged matter yesterday up to 144, today up to 176 — three sheets of what were sent yesterday were *un*corrected.

Reasons for differing the mode of lighting the grass is part of the picture.

NB Wagon is spelled in some places with two nns.

Morning Post writes unmistakable but I have altered it.

The above I noted as present over 218—232. I sent off the Clowes 2nd Revise of visit to Linyati as final. The Literary friend made a mule in one page "incapable of irrigation" I made a [?] "it would puzzle a rain maker I fear to irrigate a dispensation of providence".

6th This was stupidly put aside but I did a worse thing than leave this out. I see you answer one point and I will put Major Secard back again into the [?].

I send today 3 more sheets paged and corrected. I don't see Owen's sentence in so I put it as nearly as I remember it.

"Kakolole" is right. You will see a few more corrections I have made.

I am sincerely glad that you are pleased with what you have read. I think the Falls must please you now. I shall send you a proof as soon as I get it. They cannot alter after it is on the wood it seems [*one-third page cut away*]

The worse thing I did was writing out contents as you will see and all the while they were lying in print on my table. I sent them off on discovery but [*one-third page cut away*]

---

82. TO W.C. OSWELL

Newstead

4 April 1865

My Dear Oswell,

I have sent off the corrections you sent this morning though they are printing it. You will see if I have seen any of them. I asked of Owen's note particularly to go in. There is a clever fellow termed a "Reader" who gives a touch-up I think finally. Thanks for the trouble you took. My object in sending the 3rd Revise was that the Prof. should see it all.

Do you know any clever fellow who can tell us who tamed the Carthaginian elephant? My old Latin Professor at Glasgow can't find out. He thinks they got it from Asia. I am against Sir Roderick. I think if elephant taming be a proof of superiority then those who didn't tame them are inferior [but] he would give in [*sic*].

D.

David Livingstone.

83. TO W.C. OSWELL

7 April 1865

My Dear Oswell,

I had sent off that first Revise 97—108 as Second Revise on 3rd of this month but enclosed yours to Printer and asked Cooke who is now acting for Mr Murray to get them incorporated. About O.C. Mission Mr Murray is quite of your opinion and so am I, so we go on as we have done.

I send MacMillan to Mrs Oswell. It contains a paper by Waller which I saw in proof. Morning Post calls it "powerful", deficient in picturesqueness but valuable for its truth and earnestness. And more power to his elbow say I. [*two-thirds of page cut away*]

84. TO W.C. OSWELL

8 April 1865

Notes

Harris is my authority for *Felix Jacteata* being the "hunting pard" it is figured in his large work so.

I had sent off 3rd Revise to be paged but send this too to be mean. [?] I think Literary friend may be a she-man — Ladies write its, men *he, him*.

Panzo is the head man as you have put it.

I see you meant the Old Revise of Kebrabasa. I send what I have — I send also 201—218 First Revise and old one. I send Owen ditto so you may see it as it comes out of his hands and time be saved. You will please to forward the Mss to Clowes as soon as you have done with it. I did not say to S I would put in that to which I call your attention.

D.L

I send you some manuscript which is an antecedent step to "First Proof" you can blacken that you like and the printer has but one bill against us. Every word altered goes down against us. Every one he makes a mule of against him. I suppose our side will be ·0003, his 3,000. I meant you to look well at what I say of (page 854) Tozer = "great interests involved may be above our convictions or capacities or both" am purposely hazy. It is not all my own. I asked Stewart how he would write about him knowing how he had acted and I have of his in beginning "About the middle of 1862" and ending "to those who have not paid much attention etc." but some of the intermediate is mine too — I want you to go over it with care and alter what you don't like.

P.S.

1. Cooke tells me that paged matter is sent to you [?]
2. The old slip with Mss and all the corrections I think are inserted.
3. I send my copy to shew you all I see wrong in it.

85. TO W.C. OSWELL

Newstead

10 April 1865

My Dear Oswell,

I send today additional Mss ps 859—876 to add to the other when you send to Clowes. I think you will get a fair idea of all I wish to say about the Universities Mission. I am going on with the concluding chapter.

Mr & Mrs Webb are still in London. I shall leave as soon as they come home. I am not sure whether I ought to say anything about not selling Lady Nyassa.

[?] I would not give up the idea of doing something more by getting the. [*page cut*]

86. TO W.C. OSWELL

Newstead

14 April 1865

My Dear Oswell,

Many thanks for the service which by a note enclosed yesterday you will have found were completely successful.

The 2d page I suppose stood as it was. I don't lament having been stupid in our way of correcting etc. at first for we have now the pleasure of finding that we have improved. If we were perfect we should be sitting down like old Sandy crying because we hadn't another world to conquer.

About Sunley couldn't you put it happier? I may smooth old Towzer a little yet. The "Times" is supposed to be perfect as to printing and sentences.

Webb says that he is very sorry that he missed you — he comes here on Monday. Don't know what to do with Agnes except putting her to school again.

Tom I thought of taking to India and making him a tea planter, but I have no one to trust him with there, and I am in doubts about trusting to the chance of finding a good friend there.

With kind salutations to Mrs Oswell.

I am
Yours etc.
David Livingstone

I send nothing today. I am rather slow with something about missions which winds up the whole.

87. TO W.C. OSWELL

[Newstead]

[Date ?]

. . . part of Mss. But my wrist is numb and sore and I hastened to the clear of head and care [sic] when Webb came home they were both much better of being in London.

Would it be worthwhile making special allusion in conclusion to discoveries of Lakes Shirwa and Nyassa? I mention fever pills incidentally though I think the treatment of fever one great result of the Expedition — yet it would look quackish too — Holloway pillish!

Rowley's letter would come in well where reference is made to a Lisbon journal as I may quote him in proof of Mackenzie's disasters being the fruit of my advice. It would only be a footnote.

No signature

88. TO W.C. OSWELL

Newstead

16 April 1865

My Dear Oswell,

I sent you today the last of the manuscript 889—913 I think, with Agnes' autograph of *Finis*.[1] I laboured hard to get it all done before Mr & Mrs Webb came home and sent the Mss off after the Postman. I had a headache and it went off with the packet so I hope it won't come to you with it.

Agnes was transcribing a note for it when we had to bundle up and she ran after the letter carrier so she sends the foot-note awanting that you may put it in before it goes off to the printer.

I sent off 165—198 for paging as you recommended and 233—242 first proof for correction or rather for revise printing. I think I attended to all you mentioned as requiring my attention. I said after leaving the Lady Nyassa for a time at Johanna the Orestes towed etc. I think this is what you felt the want of. Is it not?

I put in two short paragraphs. They expect me to say something good about Mackenzie and he deserved it. Then as to disputatious fellows, I mention their way, when choking and utterly dumbfoundered by a glib tongue [of] settling the argument by "running a race", and suggest that Editors ought to do the same with their "esteemed correspondents", but you will see. The Webbs are very sorry they missed you.

David Livingstone.

P.S.
1. Meant to say that Naval men in performing duty on West Coast were not much worse of a state than in Mediterranean and the diseases on the Coast were respectable!
2. Agnes having failed to copy the letter I do it myself this morning Tuesday.

1. *Latin*: The end.

89. TO W.C. OSWELL

[*Date* ?]

Fragment

. . . would not have been half so good but for you.

You are quite right about the Portuguese — like the reformed Quaker[1] in "Uncle Tom" — I feel like "a cussing and swearin dreadful" when I think on their villainy and [?] to put it clear and mild [?] and with them slick.

Mollify if you can. Professor Owen is pleased. We will send the final proofs to you as you will see by enclosure of which I say "rejoice with me". But you know I must not make it to [*sic*] artistic for them I will not be natural and that I think a man ought always to be.

1. See Index.

90. TO W.C. OSWELL

[*Date* ?]

Fragment

I think I am right in saying Boma is the tree. Agnes wrote out the [Label] so I send it.

I enclose [Rowley] believe do you imagine it would go in last as a footnote.
Mboma a village
Boma — the tree vitex
I put out (of course) in the [?]

91. TO W.C. OSWELL

Newstead

21 April 1865

William Oswell Esq.
Groombridge
Tunbridge Wells

My Dear Oswell,
Milton's corrections seem to me very much of the "Lessons of one syllable" style "cakes of a kind the knowledge of which" or some things to that effect "they had brought back from Loanda" brought back *to them* he puts in. Then when the Town crier made proclamation long before the morning dawned or almost 1 or 2 AM he says "at early dawn". The fellows galloping without bit — bridle — saddle he says "rode back to the starting post". I said on the margin *no starting post* grandstand or Derby day have yet discovered in Africa. It seems this dilution is literary excellence if one could only believe.

I tried with might and main to make the sentence plainer about planting a stake in the ground and I could not. It made me quite nervous and at last I dashed in "Let anyone try to stick a pole in the ground by successive jerks and he will find how difficult it is to poke it into one spot"!! This I fear amounts to if you don't believe me try it yourself.

It bamboozles me fairly and I think it comes into the category of simple ideas which cannot be defined. What is a man — an animal without feathers and then the Philosopher was beat by the fellow putting down a game cock stripped of its plumage. "There is Plato's man for you."

I have asked Cooke to send me a final proof [?] after Milton.

You had better send the part about Old Bachelors marrying young wives to Owen and then you will see what he says about it. Cooke is one of them. The wretched fellow may think I am personal and not anxious for the good of the commonwealth as well.

No signature

92. TO W.C. OSWELL

Newstead

22 April 1865

Written first

My Dear Oswell,
To the best of my recollection the passage you mention was deleted and I am not aware of having repeated it. What came yesterday was sent off at once. With pen run through the underlined passages and some of the heavy parts of a double lined bit expunged too.

The part about slavery was certainly mailed to here — inserted in the body of the work and the Printers attention drawn especially to it. I send it to you now if it does not appear and you will please give it back to the Printer if it is not in.

The part it was to go in is immediately after "we heard guns firing on the opposite shore which were said to belong to Mukata" — as we returned from the Lake; or it was immediately before this paragraph and after remarks about slavery being the only trade we saw or something to that effect, and then [about] Sicard in a foot note.

"We certainly never met" — Yes There is an error in the statistics, as copied by the Times from the Blue book and I did not detect it. Thanks for pointing it out.

D.L.

The appendix with parts intended for body of work go by same post as this.

93. TO W.C. OSWELL

Newstead

24 April 1865

My Dear Oswell,
I get the part with transference of vitality and colour and put out the part about vitality and said that the colour might go.

Did because I understood that you did not like it. I should have liked [green]. I have seen it — last book. All called out against putting in [?] old ladies suckling young children, but he upheld me in it and now it is quoted by many but it does not matter.

Milton is grumpy because I put him down as the "stalking post" author. He seems to think it was you. I suppose it [incomplete]

94. TO W.C. OSWELL

Newstead

25 April 1865

My Dear Oswell,
I don't know what delays the work in London but I have all packed up and will be off by 2 today to see.

I am vexed by that fellow Milton and rather low in spirits altogether in turning up all my papers and consigning some to oblivision this morning. I lighted on the very appropriate quotation which Mrs Oswell kindly sent and which by my stupidity has been omitted. You have to bear with me. It is not intentional you know. About the "post" I walked into it on the supposition that it was Milton's. Had I imagined it was yours I would only have erased it and *MUM*. I am not sure what hotel I go to probably "Stories" 8 Dover St but will let you know. I have this morning got paged matter of the £5000 to look at but no other [one] was put right by me. "Conforms" an abomination. I think I once put in "dumbfoundered" but I have seen nothing for some time except the sheets about transference — as you did not like them I put them out.

"Being" is bad. It is not mine, nor is "comfortable" either. I thought all came back to you. Cooke assured me that Milton would never alter the sense.

I shall post this in London.
David Livingstone.

95. TO W.C. OSWELL

Storey's Hotel

28 April 1865

My Dear Oswell,
The transference of colour etc. I allowed to be struck out from a belief that you did not like it and now I don't think it worth putting back again. It was sent down to Newstead Abbey and having as if imagined your dislike to it on my mind I let it go. I would not have done it for Milton's sake.

The sentence in Owen's handwriting was sent and returned with the note "corrections all inserted — and Owen's note too"!!

Now Cooke tells me it can only go in as a table of "errata" which nobody ever reads. At least I do not. Do you?

[Whately ?] in the [?]

? 28 April 1865

Notes [to above ?]

"Planted" in exactly the same manner they plant or punch or prob or jobb the pestle into the mortar. I could not get a better word, my "Theasaurus" by Roget is at Bombay. You have the idea quite that I meant to convey.

Out with "pa".

"to once benevolence" yes — "American energy" once in is quite enough. I don't mean to repeat it.

Yes, the passage in Genesis is quite appropriate [?] put it in please. I am vexed with myself that I forgot the passage in the Psalms.

Masego and Masiko were different persons.

---

96. TO W.C. OSWELL

8 Dover St.

1 May 1865

My Dear Oswell,

On finding that the sheets did not go to you as was arranged I spoke to Mr Cooke and he at once wrote off to the Printer to that effect — I did not see them either — so things were going on anyway. I shall see Cooke about it again immediately after breakfast and shew his own note.

I went over the two last fasciculi "First proof" and put out all you indicated and some more — with a few words in addition to the point Lakes Nyassa and Shirwa, and I find it is easier to make any other corrections in slips so I sent it for a "First revise". They can then go to paging.

Cooke gave me on Saturday sheets 2a—2g as from pages 337—464 paged on coarse paper. It had come from printer to Mr Murray. I proposed to send it at once to you but he said that Printer would send you all you had not seen, so that you will have by this time from [five] sheeted up to 464 which being on coarse paper can be corrected. I see some faults and have corrected them.

After seeing Cooke I received pages 233—242 First Revise. I suppose it has come to you too for Cooke told me.

D. Livingstone

Margins
1. I don't understand how the omission of sending sheets has come about. I have been left out as well as you.
2. Nearly all. He submitted to Prof. Owen if the sheet is not destroyed.
3. The correction for "these animals" is too late — Transference is left out.
4. I looked over some sheets of Milton's and saw no important corrections in them.

97. TO W.C. OSWELL

Monday Noon

1 May 1865

I am going out this afternoon but send 233—242 in case it has not come — I don't understand why hiatus has taken place unless it has been because they (Milton & Cooke) say my [?] as to be strong in their estimation towards you and I confess that I have full confidence in your judgement and none at all in Milton's. It was rather stupid in me not to see that you tolerated the "Transference" [part] but I usually (Delete) all you don't like at once without reading the sentence again.

A bit of fun in another page. Milton put out and I put in very much because he did put it out. It is a [?] way of settling disputes which I think you have not seen — but there is no harm in it.

D.L.

P.S. I shall take a run down some day and up again. I am in for Tuesday, Wed. and Thursday.

Was at Academy dinner on Saturday & Chrystal [*sic*] Palace.

Agnes is doing it all, racing me off my feet. Poor old man. Hope your Agnes will be more merciful.

Love to Mrs Oswell.

98. TO W.C. OSWELL

Dover St.

4 May 1865

My Dear Oswell,

I went to the printer about the transference paragraph but the slip has disappeared and the matter is all out together. I tried to get the Mss but it has been in a slip and gone away. It was a loose sheet. The reason I advert to it is Mr Elwin, Editor of Pope's works and [?] Editor of the Quarterly was in Cooke's room when I went in and I told him the substance. He says he sees many reasons why it should be in and none that it should be out. Had I not stupidly mistaken your meaning it would have been right and in. I asked you yesterday to send any corrections you see in fine paper direct to me, and would further beg of your favour to put them down thus
page   6th line etc.

Read was for is

page —

page —

page —

and then make any remarks after them. I hunted up the [?] piece and erased it and will do the same by the American piece and expand a little in the second notice as my heart is softened by poor old Abe's death.

Uncanny is mine. It is my [bantling ?] a rather bastard English. Printer thinks the book very interesting. This is something I have [incomplete]

99. TO W.C. OSWELL

[date ?]

Fragment
. . . and begged that I might never see them again — Makololo. Everything up to 83 Chap XVI. I am at manuscript [?] 83 89 comes back from Clowes. I wrote telling you all I had done and lost the letter! It is time I was out of it. I hear that the Makololo are utterly squashed. All flew from each other and the black tribes and wives and children all left behind. In fact there is as we feared now no Makololo tribe.

24th. Kirk quite agrees with me. First promontory on East side of river narrow — second broad — third ditto. He is not sure about 4th, having lost all his notes. I have [?] a plan made on spot. Baines made a model but evidently never was on east side.

100. TO W.C. OSWELL

8 Dover St.

8 May 1865

My Dear Oswell,

The three [?] were too late, I am sorry to say, but I sent off at once the corrections about "midge fog", "but" and "however" and another.

I [?] unveiled — commas go as it would be to the press. The part about parting from the boat was intended to be clearer but I am sorry it is darker.

Milton gets them as the paged matter I now send and all comes to me now after he has seen it and then to Press.

I send these today and if you kindly put in the same way any corrections you may notice I shall have but one job in the incorporation. I am apt to overlook them in the page of print.

I suppose the first American part is in the paged matter now sent. I applied for them in order not to miss it.

Sir R. Murchison wants an epitome of the main objects of the book for his Anniversary address. To take a low view this is a good advertisement. What would you say were the chief points?

I hope you won't come down when Agnes and I are out, This afternoon we go to dine in country and come up again.

D.L.

101. TO W.C. OSWELL

10 May 1865

My Dear Oswell,

It is odd that the sheets sent have not arrived. I am now obliged to send off to press 2E—2H on p.417—480 with Milton's corrections which are few and unimportant so far as I can see.

I took slips to Prof. Owen — about Tozer he says that the statements are very fair and of great value to us at home and by all means to be published as they are. We altered "making no objections" to slavery to "not even protesting". I inserted the pencil corrections except "makes" because I am in [?] too about it.

After 2 months or say for at least seven months the blood of the foetus in utero circulates through the system of the parent. She mingles with her own vital fluid that of a being only half her own and through whom she becomes in a sense not usually

recognised one flesh intimately and truly with her husband. This may in additior
to simulacrity of sentiments and feelings which react on the features —
produce a likeness. [Thus] ill expressed is the idea which he did well express. He
thought that by putting transference in it might act as giving a piece of skin to a
little dog to bite and save the furniture, a tub to the whale to set the critics on a
wrong scent like — if you send any important corrections I shall still forward it
today but I am unable to detain these any longer.

David Livingstone

P.S. I have 616 pages and think I may leave out the Transference.

102. TO JAMES YOUNG

Paraffin Light Company
19 Bucklersbury
London E C

10 May 1865

. . . calling again at Mr Aston's he had made enquiries as to the amount he had and
found that he could give us £2,300 at par, and he gave me the enclosed cheque which
I place in Mr Brown's charge telling him what it is he carries for me. You will do
what you can to forward the matter. I hope Mr Hannan will make no difficulty. M.
Aston seems disposed to do everything to oblige.

I am going to call on Bevan Braithwaite this afternoon.

Meanwhile Goodbye
David Livingstone

103. TO W.C. OSWELL

11 May 186?

My Dear Oswell,
I sent off the corrections at once to the printer. The part I did not correct was abou
the decision of the argument by a race. It was not intended to be serious hence the (!

which you did not like. After "atrocious" I thought your correction made it tamer, [? *3 words*] may take it seriously then I am done for. I got three or four sheets last night [and] went over them which took me to 2 [o'clock] i.e. after the literary fund dinner. I did not mark chap. of Genesis as it would have taken time to look it up but you will do it.

D. L.

Note re Falls

"from the lips" I mean on both sides. When we performed the mad freak of going down it appeared thus [A sketch appears in the original.] "Glides" instead of *rolls swiftly* because as I have introduced the words "Then notwithstanding all the terrible dashing it has received glides gently". The depth of the thing must be enormous to admit of it becoming quiet so soon.

basaltic rock Most folks don't know that basalt is a rock. Some ladies begging their dear pardon would think it Baysalt. The first with its base on the East. Instead of First and the [?] I have added words which explain the quiet flow it so soon assumes —

Margins French, French & Music

Mrs Murray knows a very respectable Protestant [torn] I hope they will have it mildly. It is a trying time for parents but I trust all will come right.

---

104. TO JAMES YOUNG

[8/15?] May 1865

Fragment

Monday morning

I did not send off the first part of this because I received your note of the 4th bringing the happy intelligence that your complaint was not "fainting" which I did not like at all but "sickness" about the cure of which I am glad to observe some tenderness of conscience, and if you put a fee into one pocket [and] another say out of the right into the left pocket, that will [?] ample payment and you may sleep soundly and let your bed fellow sleep too.

I intend to come down on the 23—24th positively. If you are to be in London you might do your business and let us go down together. I don't smoke in the carriage nor does Agnes.

133

Will you forward the enclosed to the tailor in Glasgow, I don't know his name — the man you thought well of.

Saw Dr Stenhouse at the Royal Society — is pretty well. I had answered Nicolson [?] to the effect that I still preferred the Indian Railway and should like to know if Mr Dauncan had any other reason. I have the deed, but will see Brown first before sending my answer.

No signature.

---

105. TO W.C. OSWELL

Newstead

Monday [15 May ?] 1865

Dear O,

Great scenes of joy this morning. Webb has got a son.

I send the last sheets and wish you to look over the Mss. I must say a word in favour of Webb as you see. Alter it any way [you] see fit but it is all true. I thought of dedicating it to Webb but I can see my old friend Sir Roderick would like it and really the way he exerts himself in my favour merits the highest gratitude from me. I wish to edge in something about the Portuguese. Wishing to get our material system as on the West engrafted on the East with the world system.

D.L.

---

106. TO W.C. OSWELL

18 May 1865

My Dear Oswell,

I have just returned from saying goodbye to one of Mr Moffat's daughters (Vavasseur) and from an examination by committee of the House of Commons on West Africa.

I did but very poorly being agitated very much but I daresay they understood me.

No signature

107. TO W.C. OSWELL

Newstead Abbey

22 May 1865

My Dear Oswell,
I have dedicated to Lord Palmerston with Sir Roderick's consent and advice. It is unusual to dedicated twice to the same man. It is short and will come to you "That statesman who has ever had at heart the amelioration of the African races" —

I would like to say a word of what he has done in contrast with what has not been done on the East Coast — or in admiration of the beneficial effects of the policy which he has so long laboured to established on the coasts of Africa. Will you add a sentence to the latter effect in proof?

And also, please put an acknowledgement at the end of preface "I have [to] acknowledge the obliging readiness of Lord Russel in lending me the drawings by the artist of the Expendition which with the photography by Charles Livingstone and Dr Kirk have materially assisted in the illustrations —

Also put at the end of the Preface the date
Newstead Abbey
16th April 1865

I ask you to do this because I am off to Burnbank Road Hamilton tomorrow Tuesday evening.

Would the words "It is with sincere gratitude I thank my friends Professor Owen and Mr Oswell for many valuable hints and other aid in the preparation of the volume" grate upon your ear? Use them do and oblige me.

D.L.

P.S. Insert or erase what you choose. It is too late for the view of Shupanga.

108. TO W.C. OSWELL

Burnbank Road
Hamilton

25 May 1865

My Dear Oswell,

I ought to have told you yesterday that there was no hurry in returning the sheets because the publication is put off till Nov next — Arrowsmith came to Mr Murray about the time we were together on the 23rd and said that the map could not be done unless a man worked night and day for a month to come. I saw him 10 days ago and he said that he would go to the engraver with it that afternoon. The old heathen told Mr Murray that it was all my fault — that he had to shift all the positions 40 miles! Well Mr M thinks that upon the whole it will be better to delay it till the regular publishing time in November — The General Election is coming on and people will think more of that than of books and he knows best. I agree in his decision. Lord Shaftesbury is going to put some questions to Government in the Lords when he has read the book and though I am away it will be so noticed as to cause a sale I daresay.

A new way of courting has been tried by one of the travellers. A Baron von Huguilin who accompanied the two Dutch ladies up the Nile went into the room of the survivor a Miss Tinné with a revolver and dagger and would have her consent to marry him nolens volens[1] — she being a traveller too handed him over to the police in Cairo and saved us from electing him honorary fellow of RGS.
PS 5 m before dinner

I told you who Waller was I think — I am taking notice of the assertion made by Burton that in Africa heliometers alone count. It is made to see the missionaries. They never do so far as I can learn except when they conquer [4 *lines torn away* Kirk tells me that a party at the Anthropological use Burton as a tool and he thinks himself a leading man. They would like us to go back to the slave trade, and Lord Stanley lately talked a good deal of nonsense which will please them [5 *lines torn away*]

No signature
1. *Latin*: willy nilly.

109. TO W.C. OSWELL

Hamilton

6th June 186.

My Dear Oswell,

Your note of Friday reached on Monday as I was starting for Glasgow on business and now on Tuesday morning I say that I sent off the Preface and Dedication to

Printer all corrected — "own" "killing" and "The kindest of friends" instead of "old [?] friend. Also the errata.

I am glad that your children are getting together — I fear in the case of my mother [it] is one of breaking up. Her mind runs on poor Robert continually, and she scarcely knows me now though when I arrived she talked briskly and laughed at our little girl being disappointed that I sat writing instead of playing with my 'little sister". An attack of bronchitis was the cause of her strength declining and at her age we cannot expect recovery — Oswell takes after you — I am pressed to go tomorrow to see him walk off with prizes for German, Geography and Scripture knowledge. The number of marks it seems makes him confident of them.

Tom is rather behind in schooling from having been much absent. The Lady in Paris who takes Agnes wants her now and Agnes rebels against the heat of Paris. And I suppose I must give in. The Webbs are going to Buxton for the knee.

I was advised by Sir Bartle Frere to take [Baloochis] in my journey. Col. Rigby dissuades me so strongly that I waver. I send his letter that you may see. Is there any other class of Indians who would do — If [?] in the country could get Africans, but how to get in without a party is the difficulty — I think of trying Johanna men of the Comoro islands though they are a poor lot to trust to.

> With kind regards to Mrs Oswell
> David Livingstone.

---

110. TO W.C. OSWELL

> Burnbank Road
> Hamilton
> 9 June 1865

My Dear Oswell,

Thanks for all you have written about Indians. The chief difficulty is to get into the country with a few trusty fellows. Once in we can get good men to go from one part to another. I have written for Johanna men — Comoro Islands.

Oswell exceeded what we expected — was dux[1] in six classes and Tom got a prize for drawing — the drawing exhibited is very good. They are well pleased with your kind words. The school at which they learn is always very successful in the middle class examinations and seeing they got on I have been thinking they might prepare for trying the Indian Civil Service examination, but I don't know anything about them except that they are severe and my boys must fight their own way.

The last sheet is herewith sent. When you look over it please send to printer with ?] on it. You will see I give a parting dig at the Portuguese to fill up last page. I see the [*sic*] your way of spelling Tricolour has been changed back to Tricolor — as I had it at first. I don't know by whom but I think it is right as it stands.

Murray & Cooke think as you, we suggest to leave out Sir R's name in Dedication so I agree.

Change of air even for a single day, as for instance by going to the top of a hill is much used by common people and it certainly stops the cough when it has become a mere habit. I hope it will be successful in your [*Incomplete*]

1. *Latin*: leader.

---

111. TO JAMES YOUNG

Burnbank Road
Hamilton

20 June 1865

My Dear Young,
A telegram called me away from Oxford yesterday and by travelling all night I arrived here at 8.30 this morning. I sent the usual notice to post as not being sure if I could write a longer intimation.

My mother died on Sunday at noon — was considerably better when I left and continued so till near the change.

When that appeared Agnes said "Jesus seems to be coming for you — can you lippen[1] him". "Yes" — gave her last look to Anna Mary and said "Bonnie wee lassie" and fell asleep so gently the little thing is not in the least alarmed.

There is a peculiar solemnity connected with a touch from the grim king to one so nearly related. I trust she soon felt the touch of him who said Fear not I am he that liveth and was dead. God grant us that peace which he alone can impart.

We think of burying the body on Thursday or Friday and invite very few.

The Oxford people were excessively kind, offered me the Theatre for a lecture but I preferred a little address in Dr Daubeny's chemical classroom to a few friends.

We are all well.

Love to all at Limefield
Yours ever affectionately
David Livingstone

1. See letter no. 113 below.

112. TO JAMES YOUNG

Burnbank Road
Hamilton
24 June 1865

My Dear Young,
We performed the last duty we can observe to the departed yesterday. I proposed to go over to Limefield on Wednesday next with Agnes, Tom and Oswell, but an invitation from Mr Napier to see the trial trip of a Turkish ironclad[1] comes off on that day and it occurs to me that you and Majames would be none the worse of seeing it and then we can all go home together either that evening or next morning. If you like the trip let me know. A steamer goes from Greenock Quay to the tail of the bank about 12 o'c that day.

I have not sent to Mr Napier that I would take the young ones nor do I say it to them yet. I would like to see it myself as I never saw an ironclad working or non-working yet, and may go alone. Let us hear what you think about it.

I was very much struck last night in reading a remark of a Dr Brown in America quoted in reference to old Dr Thomas Brown [and] neglect of his body. While waiting for his death he once said *"I have worn myself out in labour which God never required of me, and for which man will never thank me"*. The underlined words are what arrested my attention.

David Livingstone.

Margin: I did not like that lassitude you had.

1. See index.

113. TO W.C. OSWELL

Burnbank Road
Hamilton
27 June 1865

My Dear Oswell,
I thank you very much for the kind note expressing your own and your wife's sympathy on the occasion of the departure of my mother. It severed many tender ties but the event had points in it which were cause for thankfulness. She was pretty well for a week. Then a change appeared only an hour before the close in a little

quicker breathing. My sister said to her "The Saviour has come for you I think, Mother you can lippen" (Scottice — trust or commit yourself to him).

She replied "Oh yes" with a tone of assent — gave her last look to our little girl (Anna Mary) and said "bonnie wee lassie" — gradually closed her eyes — gave a few long breaths and all was still.

When going away last time she said she would have liked if one of her "laddies" had laid her head in the grave and that wish has been granted. She was buried on 24th. While giving thanks now with more and more personal interest for all the dead who have died in the Lord. This was very interesting to me inasmuch as I may possess the same physical organization and the close of my career may be with the same collectedness but it is not likely I shall live long.

The Portuguese have got Lacerda's fulminations translated and published by Stanford. If so ill now, what will they be when the book comes out.

The Belchior case is fastened on — I was rather pleased with it and the continuous slavery from the Highlands to Tette ignored. I send you a copy — with love from Agnes and self to Mrs Oswell.

> I am ever affectionately
> David Livingstone

---

114. TO JAMES YOUNG

[6 July 1865 ?]

Fragment
. . . which I think handsome, and if the Portuguese refuse to put it up no great harm will be done.

I was going out to tea with, as the clavers[1] here go, with dear Mary Anne intended and my pen always more prone to nonsense than sense glided thereinto.

I want a few pounds of genuine Paraffin candles and no mixture — as they don't run like wax and other candles — nor stick together like tallow ones. I want them especially for observations. Where can I get them pure? If you come to Glasgow bring the photo-slides with you please [with ?].

> Love to you all.
> David Livingstone.

---

1. *Scots*: idle talk, gossip.

115. TO W.C. OSWELL

Newstead

24 July 1865

My Dear Oswell,

We go up to London today — Monday — and then attend Grant's wedding tomorrow and will be ready to visit you on Wednesday. We could sleep if convenient on Wednesday night and come back to London on Thursday and be ready to convoy you and Mrs Oswell down here on Saturday — so you must be ready. We are at 48 Euston Square.

E. Shelly, Sir R [?], you — Webb and I will form a real African party, so you must not disappoint us. If you think I have given too high a character to you, I am ready to abuse you on [?] as much as you like to equalize matters. The Portuguese do the abuse of me so I need no one to lower my reputation.

With love to Mrs Oswell.

I am etc.

David Livingstone.

116. TO JAMES YOUNG

[25 July 1865 ?]

. . . if less I am content.

I was at Grant's wedding today. A nice lass with a frank open countenance — £2000 a year and a London house is the tocher[1] and in addition they have the good wishes of all their friends. Speke's brother officiated.

Charles has had 2 touches of fever [in] Fernando Po — is getting a boat to go about in his consulate. He ought to get the Pioneer but wee Johnny does nothing — but I spout impudence to the Yankees. Love to Majames.

David Livingstone.

1. *Scots*: dowry.

117. TO W.C. OSWELL

Euston Square

25 July 1865

My Dear Oswell,
Agnes is out but we shall come tomorrow about 12—10, [and I] expect I can sleep on the door step or come back.

D.L.

118. TO JAMES YOUNG

48 Euston Square

8 August 1865

My Dear Young,
I have carefully considered all that you and Sir Roderick urge about remaining another month or two — it would be agreeable, but I think I shall be better employed in India trying to enlist the sympathies of the Bombay people — and see about the Lady Nyassa. Mr Tod is going to try and sell her to the Viceroy of Egypt, and she might do agreed work on the Upper Nile and Lakes, particularly Baker's if that potentate would take it into his head to put a stop to slavery there. But until I have her disposed of I have a deal on my mind.

We go on Monday next (13) to Paris to join the steamer at Marseilles on the 20th. Waller is going to see us down as far as Dover.

I don't expect salary from Wee Johnnie. I can see his reason for saying in his Aberdeen speech "that I spoke of the slave trade policy as Lord Palmerston's policy". He was angry because I had not [given] him more credit than he deserves. It will come right some how and we need not wish the little man in Heaven any sooner than is right.

Will you tell Bryson to send the Magic Lantern before Saturday next or keep it.

I have put in a Post Script to the preface as a sort of parting shot at the Portuguese which I imagine will be effective. Kirk is here (2 Lower Belgrave St) I saw him last night and today and now go off to lunch with Queen Emma at Miss Coutts.

Love to Majames
David Livingstone.

142

119. TO W.C. OSWELL

48 Euston Square

10 August 1865

My Dear Oswell,
We shall wait here Saturday forenoon till you come and will of course be delighted
to see you.

I go tomorrow to say goodbye to the old Duchess Dowager of Sutherland. I have
been getting all my luggage ready to be off and am a little tired. Parted with Mr &
Mrs Webb yesterday. God bless them. I am not demonstrative but I felt parting
with them greatly. He gracefully said "wish you were coming instead of going
away" and she shook me with both hands and shed a womanly tear. Watson was
out of town and [half page cut away] . . . that I dined with Queen Emma last night,
a nice sensible person and none stuck up as yet she seems quite a superior woman
[half page cut away].

120. TO JAMES YOUNG

48 Euston Square

12 August 1865

My Dear Young,
I must now say farewell and wish blessings on you and yours according to your need
from the ever present Friend.

Many thanks for all your kindness to me and mine.

We start on Monday at 10 from Charing Cross for Dover etc.

I enclose a statement of accounts from Mr Murray at his desire after I had looked
at it. He will pay you the amount.

Letters and papers sent to Mr Lennox Conyngham will be forwarded.

Mr Wylde also of the Foreign Office will send letters. He is a better fellow as Kirk
will tell you than some of the others.

Agnes and I write in kind love to Majames and all the family.

Ever Affectionately
David Livingstone.

121. TO W.C. OSWELL

Marseilles

19 Aug 1865

My Dear Oswell,

I think Agnes is all right. Revd M. Calliatte the father of her instructor Madame Hocédé is a nice old protestant minister and rather clever as a writer of reviews in the "Revue des deux Mondes" in association with M. Guizot, and he lives principally in a village near Dreux called Marsauceux. Agnes is with them there for a month or so during the vintage. Mrs Callliatte is Miss Lemue whom you remember at Motito. Mr Moffat lived with them a week when he was in France and all their belongings are religious people and thoroughly Protestant. Mrs Hocédé is about 30 — lost her husband after 11 weeks of married life by scarlet fever when she was only 19, seems clever — is certainly a good musician and Agnes likes her. The country is a wine growing one. The people well off and all agricultural. Dreux an old town contains burying place of the Orleans family. The whole country is on the chalk dry and healthy. She goes to Paris in a month or so. Their two girls are to join her, and she will be on visiting terms with some of the best Protestant families as M. Triquety the artist. Poor thing she was very much cut up by parting with me. She was much attached to poor Bob whose name I fear must be heard no more — and I felt rather alone in the world. Hold her. I told her to write to you in any case in which she required counsel as I am sure that you would give her wise hints for her guidance. We sail hence tomorrow 20th. God bless you all.

Ever affectionately yours

David Livingstone

Agnes' present address is
at Revd M. Calliatte
Marsauceux
Près Dreux
Eure et Loire
France

PART FOUR
The Last Journey
1866 • 1873
CHRONOLOGY

1866	April	Beginning of journey inland
	August	Arrival at Lake Malawi
	September	Westwards across south end of Lake; desertion of Comorans
	December	Luangwa escarpment
1867	January	Medicines stolen
	April	At south of Lake Liemba (Tanganyika) seriously ill
	November	At Lake Mweru; at Mwata Kazembe
	December	Delayed by slowness of Zanzibari caravan
1868	July	At Lake Bangweulu
	September	Journey to Ujiji
1869	January	Ill with pneumonia
	February	West coast of Lake Tanganyika
	March	Arrival at Ujiji; stores sent from coast found stolen
	July	Across Lake Tanganyika to 'Manyuema country'
	December	Abortive trip to Lualaba river
1870		Return from Lualaba to Bambarre, seriously ill with ulcers.
1871	January	Arrival of porters sent by Kirk from Zanzibar
	March	Short fruitless trip on Lualaba
	July	Nyangwe massacre; departure
	October	Return to Ujiji, stores again found stolen
	27 October	Arrival of H.M. Stanley (date as calculated by Mr I.C. Cunningham); 'Doctor Livingstone, I presume'.
	November	Visit to north of Lake with Stanley

1872	March	At Unyanyembe (Tabora) with Stanley Departure of Stanley
	August	Arrival of stores from coast; departure from Unyanyembe southwards
1873	1 May	Death at Chitambo's village near Lake Bangweulu; embalmed body carried to coast and Zanzibar
1874	18 April	State funeral in London; entombment in Westminster Abbey
1874		*Last Journals*, edited by H. Waller, published

At the end of Narrative of an Expedition to the Zambezi . . ., *published when Livingstone had already reached Bombay, he declared that an effective way to end the east African slave trade would be 'an expedition or settlement inland'. Elsewhere at this time, he stated that purely geographical questions did not interest him, and that he intended to set up a trading station in Africa — but his official sponsors in Britain had commissioned him to delineate the central African watershed and establish whether it was the source of the Nile. With these varied objectives, the Livingstone party landed near the mouth of the Rovuma, north of any Portuguese settlement, a few days after leaving Zanzibar. With him were sixty porters, including Susi and Amoda, Chuma and Wakatini, and Joseph Gardner, another freed slave who had been educated at Nassik in India. There were twelve Indian soldiers and their officer, and men recruited from the Comoros or hired on the spot. There were also camels, donkeys and buffaloes. But although Livingstone was armed with a British consulship and a passport from the Sultan of Zanzibar, the expedition was both undercapitalised and underequipped.*

Livingstone's initial plan was to go around the north of Lake Malawi to see if it were linked by water with Lake Tanganyika, but the activities of Zanzibari slave raiders, and of Nguni-Zulu marauders, called the Mazitu by Livingstone, made this impracticable, so he skirted Lake Malawi to the south and pressed on to the north-west. The Indian soldiers had proved unsatisfactory, and Livingstone sent them home. Shortly after passing the Lake, the Comorans deserted him. (When they reached Zanzibar they said that he had been killed by the Mazitu and when this news reached Britain, the RGS persuaded the government to spend £1200 on a search expedition.) Wakatini, one of the freed slaves, found his people again, and stayed behind in his village.

Thus depleted, the party crossed the Luangwa and Chambeshi rivers and reached the southern end of Lake Tanganyika. Livingstone's medicine chest was stolen on the way, and near the Lake illness and shortage of supplies forced him to link up with Zanzibari slave and ivory traders. With them he travelled to Lake Mweru, to Mwata Kazembe's capital, and then north to Ujiji, after making a separate journey to see Lake Bangweulu. During this period he began to believe that the Nile rose on the central watershed, which he had identified in the high ground between the Luangwa and Chambeshi rivers.

At Ujiji he found that most of the stores he had ordered from Zanzibar had been stolen, but despite this, and despite being left with only a remnant of his party as a result of desertions, he travelled west, back across Lake Tanganyika, again in the company of slave and ivory merchants who were moving into a new area of depredation. Livingstone hoped to reach the large river which flows north out of Lake Mweru, and to establish whether it was the headwaters of the Nile.

The part of Africa which Livingstone had entered, and which he called Manyuema, was in a state of confusion, for reasons mentioned above, and his attempts to examine the large river, the Lualaba, were frustrated. He was also short of supplies and trade goods, and fell ill for many months with tropical ulcers, which he treated with powdered malachite. While he was recovering, he was joined by porters sent from Zanzibar by Kirk, but they had left his stores at Ujiji. He

The Last Journey

would have continued his researches, however, had it not been for the massacre at Nyangwe, when Zanzibari slavers opened fire indiscriminately into the crowded market and caused the death of hundreds of people. The Nyangwe killings decided him to return. (When his report of the incident reached London later, British public opinion was so outraged that the government found a moral reason to force the Sultan to close Zanzibar's slave market.)

When Livingstone arrived again at Ujiji, he found that the stores brought from the coast by Kirk's porters, had, like the previous consignment, been stolen, and he was saved from destitution only by the timely advent of Henry Morton Stanley with an abundance of supplies. Stanley was a journalist who had been commissioned by his newspaper, the New York Herald, to 'Find Livingstone', from whom nothing had been heard in years as Zanzibari merchants were reluctant and often totally unwilling to take his letters to the coast.

Stanley stayed with Livingstone for four months, during which time they made an excursion around the north of Lake Tanganyika (which disproved Burton's theory that it was the source of the Nile) and when he returned to the coast he took Livingstone's journal, letters, and despatches with him. He tried to persuade Livingstone himself to go with him so he could rest and get medical treatment, but this was refused, with the request only that Stanley send supplies and porters from Zanzibar. When Stanley reached Bagamoyo, he found a British government expedition setting out to do what he had just accomplished, to find Livingstone; this expedition, which included young Oswell Livingstone, turned back. Before Livingstone died in 1873, two further search expeditions set out from Britain, one of them official, the other paid for by James Young.

From Zanzibar, Stanley sent the convoy of porters and stores Livingstone had asked for and sailed for Britain, where his reports, of his own adventures and of Livingstone, brought both men fame and glory. Livingstone's reputation was restored to the heights it had enjoyed after his first great journey, but of this he would remain unaware.

When Stanley's supplies arrived in the interior, Livingstone set off on his final travels. He intended to go around the south of Lake Bangweulu, visit the presumed 'fountains of Herodotus' and the copper mines of Katanga, then down the Lualaba to establish once and for all whether it were the Nile. Then he would return to Britain, where he had asked a friend to find him lodgings in London.

Livingstone and his party, which with Stanley's reinforcements now numbered over sixty, flanked Lake Tanganyika on the east, crossed northern Zambia, and after a nightmarish struggle through the Bangweulu swamps crossed the Chambeshi and reached the village of Chitambo to the south of the Lake. Here Livingstone died on 1 May 1873, a few weeks after his sixtieth birthday.

Susi and Chuma took command of the caravan. Livingstone's heart and viscera were buried with a Christian service near Chitambo's village, the body was embalmed, wrapped in cloth and carried to Bagamoyo together with Livingstone's books and papers. The 1600 km journey took eight months, and during it ten members of the party died. When the body and Livingstone's effects were given to

*the British authorities, Susi, Chuma, Amoda, Gardner and the other carriers were
paid their wages and summarily dismissed.*

*The body was shipped to Britain, and after being subjected to a post-mortem
examination to establish that it was in fact Livingstone's, was given a public funeral
and entombed in Westminster Abbey, the last resting place of many British
heroes.*

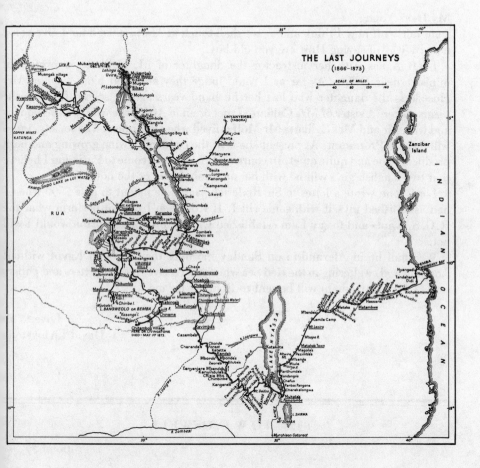

THE LAST JOURNEYS
(1866-1873)

SCALE OF MILES

122. TO JAMES YOUNG

On board SS Massilia
22 August 1865

My Dear Young,
I am fairly off now I think and as we shall touch at Malta in an hour I give you a note by way of saying How are you old boy.

I left Agnes with her instructor the daughter of M. Calliatte, a protestant minister near Dreux. As far as I could judge they are good pious people. Mrs Hocèdé is the daughter who lost her husband years ago by scarlet fever — she seems clever. A sister of Mrs Calliatte I met often in Africa the wife of a missionary and with Mr and Mrs Calliatte Mr Moffat lived when in France. The society Agnes will meet is Protestant. At present she is in the country, a wine growing and corn producing one and quite quiet. In course of time she will come to Paris and I believe that two English girls will be with her as coinhabiters of the house [?].

Lord John wrote a letter to Sir Roderick saying he meant to give £500 when I had established myself with some chief. If I do then I cannot perform what the R.G.S. wants and to say I am established when I am not as Burton would I will not.

We shall be in Alexandria on Sunday morning then through Egypt without stopping and sweltering in the Red Sea when you receive this — Letters and papers sent to Mr Conyngham will be sent to Bombay for me.

Love to all
David Livingstone

123. TO W.C. OSWELL

Bombay
29 Sept 1865

My Dear Oswell,
As I may not have leizure when nearer my starting point I write a note by this mail to say that I am getting on pretty well in the way of preparation for leaving in November. I have got eight Africans some of whom have a knowledge of carpentry and smith work and they may be interesting if we try to build up on a canoe for navigating Tanganyika. I am to get some men of the Marine Batallion who have roughed it already and can manage buffaloes. The Govt ensures their pay, pensions, allowances as if on actual duty for the State. His Excellency is a first rate

man and enters into my project with great heartiness and goodwill. The reason why I wish to take some tame buffaloes is I think there is a chance of them being able to withstand the poison of the Tsetse. They are wonderfully like the wild Nyaris I have seen horns with the genuine curves. They are surely more than half brothers though the males have not the horny foreheard. I [intend] them to be beasts of burden. If they stand the tsetse then we shall be conferring a boon.

The Sultan of Zanzibar is expected here on a visit on the 5th Oct so I shall confer with him as to getting supplies by way of Quilloa. The Governor is going to treat him with distinction by way of giving him some new ideas — will show him all now doing in Bombay.

Col. Playfair the resident at Zanzibar has left ill of heart disease. I wish Kirk could get that post. He would be invaluable there but we have no influence. I am going to sell Lady Nyassa if I can. I suppose I shall not get more than £2000 but it's all right, I tried to do good though I failed. The Governor thinks that it is a considerable step in advance to get the Portuguese to be angry and write by Lacerda — a few years ago they would not have minded.

No signature

---

124. TO J.S. WILLANS, ROYAL ENGINEERS

At Remington and Co.
Bombay

16 November 1865

My Dear Sir,
I am sorry that from having neglected to bring away the folded map of Africa by Arrowsmith I am obliged to trouble you to send it to the above address at your convenience.

I have delayed till now because I proposed to visit Poonah but that being now doubtful I hope you will excuse my troubling you with this request. Possibly you may know some friend who will take charge of it.

I never had an opportunity of thanking you personally for the large map which you kindly undertook to get constructed [but] now do so heartily. It answered the purpose very well indeed.

Sincerely yours
David Livingstone

125. TO W. SUNLEY

Bombay
31 December 1865

My Dear Sunley,
I got your letter yesterday and will feel much obliged if you can engage six good boatmen at Five dollars pr month the sixth being headman to have more. The Wasp possibly may go to your island before going to Zanzibar and the commodore having very kindly offered to take anything for me by her I write to you just in the chance of her going your way first. I have no doubt you will do what you can and any advance you may give please send the account to Dr Seward for me. If a dhow were coming to Zanzibar before February that would do to bring them.

I have thirteen native carriers and eight Africans but they are not boatmen equal to the Johanese — Alla Maria my former headman would have 10 dollars a month, any untried man might do with seven. You have heard of Rae's sad end — married, ill in bed one day with perforation of stomach and dead next morning.

Kirk is well and I think coming this week but I have no time to write.

Ever yours
David Livingstone

126. TO W.C. OSWELL

Bombay
1 Jany 1866

My Dear Oswell,
Many happy returns of this day to you and yours and a blessing from above on them all I was very much delighted with your letter of the 2nd Novr I got the Saturday and Atheneum Reviews which are favourable enough perhaps too much so — My old antagonist John Crawford of the Examiner not quite so much so but still fair and the Reader by Winwood Reade enough to make me shut my eyes; if it touches up the Portuguese to change their infamous system I shall be content —

You would see a paragraph about Van Der Decken having got into difficulties on the Juba — he must have been going on too fast — knocked two holes in his steamer's bottom went ashore and a party came down and attacked the vessel killing some. His Prussian Lieutenant of the Navy skedaddled or something like it & it is not known whether he survives or not — His great object was to prove Speke wrong — When Thornton was with him he sent up a rocket or two every night to ensure respect!

Dr & Mrs Beke were at Suez lately going up to implore King Theodorus to liberate the captives — Mrs Beke had cut her hair to don the male attire, as if she didnt wear the breeches already. It will be curious if a dumpy little woman can pass among savages as a man. Lady Franklin is here to spend the pleasant months in India for health & perhaps pleasure.

These are our local news and I may add Kirk has got an appointment at Zanzibar as I believe Assistant Political Agent. I am glad of this for he is a good fellow in every way — he was telegraphed to yesterday by the Governor who said to me it was a great recommendation to him that I was anxious that he should be there I am to start on the 3rd I am all ready & very tired at being idle — the Thule is one of Sherard Osborne's late fleet & is going as a present from this Government to the Sultan of Zanzibar. I am a free passenger and have the honour to make the formal presentation — this will give me a little lift in the eyes of the Arabs. I take two mules as a sort of experiment too in case the buffaloes have not done well at Zanzibar. I have been thinking that a settlement of the Sultan at the mouth of the Rovuma — if they could only be persuaded to make it a free one might be a step in advance but few if any of his people can see that free labour and free trade would in the end prove the most profitable — They would stand the malaria better than Europeans for we whites will do the most foolish things possible in the most unhealthy localities. I think Kirk will do what he can to promote freedom and commercial intercourse.

Waller tells me that the friends of Tozer hang down their heads in dismay at what I have said, some aver that a man should not be spoken against in his absence. Why didnt they teach Tozer that first?

Baker married his mistress at Cairo and from all accounts she deserved it after going through all she did for him. I heard about his woman but it was not made public and if she turns out well better that it never should; he has a good smack of negrophobia and it is amusing to find him pronouncing on the absence of all knowledge of a supreme being among a people of whose language I suppose he knew as much as Gordon [Cumming ?] who had only about six words — Nyama pololiolo — Chukuru — & Aretsamae.

Nothing of Robert after his capture by the confederates — he was so fond of Agnes he would have written to her had he been alive — it would be better if we knew but I fear we never shall. Baines' book contains very little. Old Crawford says his description of Victoria Falls is better than ours — that's a fib — I say ours for without your aid the critics would have had lots of faults to find. My love to Mrs Oswell & the bairns also to Mrs Vardon.

From your ever affectionately
David Livingstone

Margin: Lennox Conyngham — F.O. is a good way of sending anything — It comes to the Governor of Bombay in an official envelope — thanks for remembering Agnes and letting her know she is not forgotten.

153

127. TO JAMES YOUNG

2 Jan. 1866

. . . Van der Decken has come to grief I fear. He went on too fast knocked two holes in the bottom of his steamer — was attacked while he was ashore and his Leftt a Prussian naval officer skedaddled — when with Thornton he used to let off sky rockets every night by way of securing respect ! !

Lady Franklin is here enjoying the fine mild weather — why does not Lady Young take her young man out [?] a trip — she does not plague enough I think — my love to her and all the children. Tell me all about them my dear fellow. It is always interesting to hear about those we love — like — you Scotch bodies would say. I say love cause I am African. Is the old [*batchelor freethinking*?] himself yet and mending his ways — Love to Mary and them all.

I am off tomorrow morning. I have two mules in case of the buffaloes failing. Dr Beke and his wife are going to Abyssinia to implore King Theodore to liberate the prisoners. Her hair is cut in order to assume male attire as if she did not wear the breeks[1] already — she is about Majames size in diameter — fancy Majames in male attire. It would make me swarf[2] to see it.

I am in great hurry and you will excuse shortness and not make yours shorter. Lenox Conyngham will forward all you write to

Yours ever affectionately.
David Livingstone

1. *Scots*: trousers.
2. *Scots*: swoon.

128. TO G. FRERE

Zanzibar
7th March 1866

My Dear Mr Frere,

Before passing away to the Interior again I shall just say "Howdye do" and thank Mr Layard and you for the kind way in which you dealt with a scandal which did but little credit to its soft headed propagator I refer to your letter of 19th May 1864 which I saw only two days ago. Bless the fellow! ! all lay beached at Mosambique for some six weeks and the people we were falsely said to have "taken out with closed hatches" found two women [?] who had been brought down from

154

Magomero with 500 others, and sold under the eye of the immaculate Governor-General de Almeida, but if I had been weak enough to act the slave trader was it not a pity. Did I not deserve the exercise of that "most excellent gift of charity". Waller wrote an indignant protest to the bishop, but he ought to have sent it to you.

In the slave market here I saw at different times from 70 to 300 slaves exposed at once for sale. Northern Arabs and Persians were the chief purchasers. They have their dhows in the bay or southwards picking up what they can in slaves from Portuguese and Malagasse. They cannot go North till April or May partly because the exportation coastwise is prohibited, but chiefly because the wind won't let either Northern or Zanzibar dhows go North. The prohibition, you see, is exactly in accordance with the laws of nature. We must never go contrary to the laws of nature, this is modern philosophy. As soon as the monsoon changes all the dhows will crowd North. Then we have no prohibition. This is true philosophy but not slave trade suppression policy.

The slaves swarm here and the majority are Manganja — their prices when I was in the market were from 7 to 20 dollars. I see little chance of their lot ever becoming better — so long as they are about on a level with their Arab masters and will do very well. Master and man partake of the general indolence and barbarism but with the advance of civilization or trade and luxury their lot becomes harder. The lust of gain in the master must always increase the hardships of the slave — so if we wish well to the slave, we must wish the Arabs to remain in a state of barbarism. We must no interfere with the status of slavery, however, if I might venture to say it we perpetrate a monstrous mistake in allowing slaves to be carried coastwise or anywise mortal not to injure the status of slavery here. They virtually uphold it. Why should that which is piracy everywhere else be allowed between this and the mainland? Answer, because we have no right to touch the status of slavery — It is said that without a fresh importation of slaves into Pemba and Zanzibar these islands would soon become depopulated, to people these wretched sinks we allow the depopulation of hundreds of square miles of the Manganja and Lake Nyassa country. It makes me sick at heart to think of it, and raises bitter regret that we did not work out our experiment of cutting up slaving at its source.

With kind regards to Mr Layard.
I am &c
David Livingstone

129. TO —

[Rovuma ?]

20 May 1866

. . . The Havildar is hereby ordered to come up to me on receipt of this and to give evidence as to the disobedience of orders by the sepoys under him. He is to inform them that as they are reported to have refused to get up in the mornings to march after repeated orders by the Havildar — or to carry their muskets and belts, or cook their own food — I shall take this evidence in order [to] report to the Governor of Bombay — I shall take the evidence of the Nassick boy too, add my own which I send — They left their duty whenever I was out of sight, sat down and smoked and ate in the march. Their sickness was not fever but eating too much, and vomiting and all the people of the country who saw their skulking when my back was turned said *"Your men are not good"*.

The Sepoys are to remain where they are till I send orders to them to march [?] to the coast and to Bombay as prisoners to answer the evidence given against them — I meant them to know exactly the [charges] they will have to answer . . .

130. TO A. SEDGWICK

Lake Nyassa

24 August 1866

My very Dear Friend

I have taken a sore longing to write to you though I have not the faintest prospect of being able to send a letter to the sea coast. The Arab slave traders on their way thither avoid me as if I had the plague — In six or seven cases they set off across country as soon as they heard that the English were coming and dashing through bush and brake[1] here is a more serious affair than with you for the grass is generally over one's head in the hollows — as thick as a quill in the stalk and often intertwined with creepers — I would fain have spared the slaves who were thus dragged, but the masters took care not to look me in the face — one sagacious old leader who had about 800 slaves in his party, hearing that, after a march of eight days through a fine country completely depopulated by the slave trade, we were nearly famished, and that we were just at hand, came forward and presented a bag of flour and an ox — I daresay he had some genuine goodness in him though it looked like taking the "bull by the horns". I had only three or four of the strongest of our party, and we were making a forced march in order to purchase and send back provisions to the weaker still behind, but this was the only chance we had of sending a letter, and our friend could not wait till we had written — He was beginning that long march which we had just finished, and every hour to him was precious.

To give you an idea of the country it is a gradual slope from the coast up to within forty or fifty miles of this Lake — The first 80 miles or so are covered with dense forest, the only bare spaces being the clearings of the inhabitants — The rock where it can be seen is coarse grey sandstone with blocks of salicified [sic] wood lying on it. This overlies coal. Beyond the sandstone we come on gneiss and sometimes granite, there the forest is scraggy but it is still so thickly planted that one can rarely see the horizon — About 200 miles inland the country becomes undulating and on the crests of the waves one sees mountains all around — great rounded granite masses — igneous rocks appear among these masses and large patches of ferruginous conglomerate are met with — The country is still rich in soil but the trees are small as we attain greater altitude — and the number of running rills in the mountainous district is quite astonishing — I counted in one day's march no fewer than fifteen flowing burns[2] — These are the sources of the Rovuma — The water shed between it and the Lake rises up to 3400 and even 4000 feet. This elevated region is just Magomero magnified, and to this poor bishop Mackenzie hoped to extend his mission — I had to wear my thickest flannels — The water though only 61 degrees felt much too cold to bathe in — cattle shewed that no tsetse exists, and large patches of English peas in full bearing shewed how English vegetables could flourish. The most influential chief on these highlands is Mataka of Ajawa or Waiyau extraction — some of his people had without his knowledge gone to the Lake and carried off a number of cattle and people — Mataka had ordered them to be taken back before we arrived, but I accidentally saw the party — The women and children numbered 54. The young men and boys about a dozen, but they were then employed among the cattle which were about 30 head in all — This spontaneous act was the more creditable in as much as he has been subjected to none but slave trading influences, and all through this region the process which bishop Mackenzie stemmed goes on annually — The Ajawa or Waiyau make the forays, and the Arabs furnish the powder and guns — I look at this fine region fast becoming depopulated with feelings of inexpressible sadness — We allow the Zanzibar slave trade within certain limits, and the effect of this license is, that insignificant island is a great slave emporium and hundreds of miles of a far finer country is annually swept of crowds of people — our mission there is virtually gagged — The Sultan is all civility and flattery but no missionary progress can be allowed among his bigotted Mahometans and then about half the missionary strength must always be absent in quest of health — It is almost enough to make Mackenzie turn round in his grave to find his mission degraded to a mere chaplaincy of a consulate — and I fear that there is no hope of seeing Central Africa occupied by its mission in our day — I was very much delayed by wanting provisions, and by the laziness of some sepoys whom I had to dismiss but the easy boating of about three weeks to Magomero will bear no comparison to the four months hard toil we had in coming here — When we see how bigotted and unfriendly the coast tribes are and how friendly the people in the Interior prove themselves I conclude that Africa must be christianized from within. Where ever a path may be found I can conceive of none superior to that by the Shire — It is true we lost valuable lives — The Arab who proved our friend in need told me as a piece of news that the Quilloa slave traders had lost one hundred of their number in one sickly year. I saw seven of their graves and surely the church can afford to spend lives in saving as freely as others can in destroying — But after all I have hopes in the church yet and I cannot believe that all the hopes, prayers and sympathies which clustered around my poor Dear friend are to go into thin air. I ought to have mentioned that on the highest part of the watershed we had mica schist and then when we come down here the gneiss is thrust away from the Lake and tilted right on edge — The strata often dipping quite perpendicularly but I can see no agent that thus thrust it away — It felt like coming to an old home to see Nyassa again and dash in the rollers of its delicious waters — I was quite exhilarated by the roar of the inland sea — The people have generally been friendly with us — Though from their intercourse with coast Arabs they have learned some of their ways suspicion has not taken root. I have not yet learned that any Arab had endeavoured to propagate Mahometanism — Their zeal for that faith seems to have perished and without being uncharitable I don't think they have much of it themselves — I have been rather surprised at finding on undoubted testimony that even where Polygamy has free scope the venereal disease is rampant. Two English medical officers assured me that they had seen nothing like its prevalence in Europe — One of these officers was so scandalized on finding what we call respectable married men with it that he threatened to complain to the Sultan — This says little for Polygamy.

I have several times recollected a remark made by the Dean of Ely in your house that he might be able to do something to promote the education of my children. I did not think much about it at the time but it has since struck me that if I had the opportunity I would tell him that I shall esteem it a great kindness if he in any way remembers them — I am rather oddly situated as to friends — I have a great many sincere true hearted ones for whose favour I feel very grateful — To them for their favour and to Him who disposes their hearts to feel kindly towards me, but again and again I have been left in the lurch — one mentions some new and most interesting book "I would have send it" he adds "but you have so many friends I am sure some one must have sent it" so with news. They are sure some of my "many friends" must have given all. As he mentioned education spontaneously it seemed as if he were not one of the many who feel that others are so much more liberal than themselves — I have one son at Glasgow College sixteen years of age — Another at a private school in Hamilton about ten years of age. Should the Dean wish any information about them, Professor Andrew Buchanan of Glasgow would supply it or James Young — Limefield — West Calder — Scotland — another of their guardians. The Dean may have nothing at his disposal but I do not value his kindly feelings the less and I am sure that you will excuse my asking you to give the above information at your leizure.

I think of you as I saw you last at Norwich and beg you to present my salaam to Marybell and to any of your Bouquet of flowers you may meet. I leave a little space to fill up when I see a prospect of sending a letter.

Bemba Lat. 10°10′ South Long. 31°50′ East — 1st Feby 1867. I have been a long time in working up to what is probably the watershed I seek — 4500 feet above the sea and the Loapula in front — A hungry time we had in passing through the dripping forests of the Babisa country — no animals to be shot and the people had no grain to sell — Mushrooms in plenty though but, woe's me, good only for exciting dreams of the roast beef of byegone days — no salt either. This causes the gnawing sensation to be ceaseless but we got through by God's great mercy — sugar we have forgotten all about and roast a little grain to make believe it is coffee but we have got to a land of plenty and are going to have our christmas feast tomorrow. We had nothing to celebrate it when the day passed but won't be balked of it for all that — I am excessively lean but take on fat kindly as do some races of pigs — I have heard nothing of home since we left the coast but pray that the Most High may keep all my dear friends and relatives. Among [whom] I am glad to place my very Dear

Professor A. Sedgwick
from
David Livingstone

Margin: It is possible that you may know Mr Young by report — formerly my teacher in chemistry — has made a fortune by Paraffin oil and is a fine straight forward good man.

1. *Scots*: scrubland.
2. *Scots*: streams.

---

131. TO J. YOUNG

Country of the Chipeta

10 Nov 1866

Private

My Dear Young,

I have felt anxious several times because could not advise you about £400 which with £600 of my own I left at interest with Messrs Coutts — and which £400 with its portion of interest I wish to be transmitted to Zanzibar to be available for the payment of wages and other things on my return. It would be best for me in gold and perhaps Messrs Fleming & Co would send it to Captain Fraser or whoever may be managing their affairs at the island when you receive this.

The Last Journey

Messrs Coutts require some little notice to be given of an intention to draw out what is placed at interest, and I think that shewing this part of the letter may be taken by them as sufficient authority for drawing the £400 and odds.

It has been quite impossible to send a letter coastwise ever since we left the Rovuma. The Arab slave traders take to their heels as soon as they hear that the English are on the road. I am a perfect bugbear to them. Eight parties thus skeddadled and last of all my Johanna men frighted out of their wits by stories told them by a member of a ninth party who had been plundered of his slaves, walked off and left me to face the terrible Mazitu with nine Nassick boys.

The fear which the English name has struck into the souls of the slave traders has thus been an inconvenience. Could not go round the North end of the Lake for fear that my Johanna men at sight of danger would do there what they actually did at the Southern end, and the owners of two dhows now on the Lake kept them out of sight lest I should burn them as slavers and I could not cross in the middle.

Rounding the Southern end we got up Kirk's range, and among Manganja not yet made slave sellers. This was a great treat for like all who have not been contaminated by that blight they were very kind, and having been worried enough by unwilling sepoys and cowardly Johanna men, I followed my bent by easy marches among friendly generous people to [whom] I tried to impart some new ideas in return for their hospitality.

The country is elevated and the climate cool. One of the wonders told us in successive villages was that we slept without fires. The boys having blankets did not need fire while the inhabitants being scantily clad have their huts plastered inside and out — even the roofs — to make themselves comfortable. Our progress since has been slow from other and less agreeable causes — some parts have been denuded of food by marauding Mazitu or Zulus. We have been fain to avoid these and gone zigzag. Once we nearly walked into the hands of a party and several times we have been detained by rumours of the enemy in front.

January 1867 I mention several causes of delay. [I] Must add the rainy season as more potent than all except hunger. In passing through the Babisa country we found that food was not [to] be had. The Babisa are great slave traders and have in consequence but little industry. This seems to be the chief cause of their having no food to spare. The rains too are more copious than I ever saw them anywhere in Africa. But we shall get on in time; I often think of you all and many of your acts of kindness assume during the process of being turned over more of their due proportions than before. All the trouble you have been at with me and mine — the expenses so cheerfully paid of that delightful Highland trip by Lochfine and the other by Lochearn Head and many others come back vividly before the mind, and I bless Him who has given me such a friend.

The material aid was great but the affection which evidently prompted you and Mrs Young was beyond price — invaluable. I fear I never thanked you enough but your kindness would excuse me.

I am sorry that I never could write to Dr Stenhouse about his invention. The sheet his agent gave me to place on the ground beneath my bed has been invaluable as a tent overhead. He offered me a covering of a lighter kind and I regret exceedingly not having accepted it. The Mackintosh sheets I have tried are not to be mentioned in comparison. This black sheet is lighter and lasts wonderfully while the India rubber sheet so glues itself together that you soon tear it to pieces in drawing the folds asunder. The first pair of shoes have lasted during a five hundred mile tramp, often over tough stoney soil and in the driest hottest season. I gave away the first pair not because the uppers were broken or the soles worn out but because the inner seam had given away at the toes and the heels were gone. I ought to have had a pair *not* Stenhouse to try against the others.

I am now putting a second pair to a severe test — daily wet outside and in and then exposed to a broiling sun. If they last long at this, I shall let the Dr know. I think his invention really very valuable, and I wish you would give him this extract as a sort of acknowledgement for kindly providing that *brick* of a sheet.

1st February. I am in Bemba or Lobemba and at the chief man's place, which has three stockades around it and a deep dry ditch round the inner one. He seems a fine fellow and gave us a cow to slaughter on our arrival yesterday. We are going to hold a Christmas feast off it tomorrow as I promised the boys a blow out when we came to a place of plenty.

We have had precious hard times and I would not complain if it had not been gnawing hunger for many a day, and our bones sticking through as if they would burst the skin. When we were in a part where game abounded I filled the pot with a first rate rifle given me by Captain Fraser, but elsewhere we had but very short rations of a species of millet called [Maere ?][1] which passes the stomach almost unchanged.

The sorest grief of all was the loss of the medicine box which your friend at Apothecaries' Hall so kindly fitted up. All other things I divided among the bundles so that if one or two were lost we should not be rendered destitute of such articles, but this I gave to a steady boy and trusted him. He exchanged for a march with two volunteers who behaved remarkably well till at last hungry marches through dripping forests, cold hungry nights and fatiguing days overcame their virtue and they made off with "steady's" load. All his clothes, our plates, dishes, much of our powder, 2 guns, and it was impossible to trace them after the first drenching shower which fell immediately after [they] left us. The forests are so dense and leafy one cannot see fifty yards on any side.

This loss with all our medicine fell on my heart like a sentence of death by fever as was the case with poor bishop Mackenzie — but I shall try native medicines and trust in Him who has led me hitherto to help me still.

We have been mostly on elevated land between 3000 and 5000 feet above the sea. I think we are now on the watershed for which I was to seek. We are 4500 feet above the sea level and will begin to descend when we go.

This may be put down as 10° 10′ South Lat. and Long. 31° 50′ E. We found a party of black half caste Arab slavers here and one promises to take letters to Zanzibar, but they give me only half a day to write, but I shall send what I can and hope that they will be as good as their word. We have not had a single difficulty with the people but we have been very slow — eight miles a day is a good march for us, loaded as the boys are, and we have often been obliged to zigzag as I mentioned.

The Last Journey

Blessing on you all. Love to Mrs Young from yours ever affectionately

David Livingstone

(PS) I have opened this to add [?] for reading over a letter I wrote some time since to Professor Sedgwick I had forgotten.

The Dean of Ely Revd Dr Goodwin Harvey said to me that he might be able to do something to aid in the education of my children. Mr Hannan you remember said more and publicly too. Now these English people have much in their power in their richly endowed schools and colleges, and as the remark was quite spontaneous he may do something. Should Tom's desires tend to greater advances in education than he can have in Scotland he may be placed on some foundation as they call it. I refer the Dean to you and Dr Buchanan for information if he wishes it about either Tom or Oswell. You will use your own judgement and Dr Buchanan's. I cannot possibly write the Dr though since their sore afflictions my heart feels knit to them all more closely than ever.

A hearty salaam to friend Bartholomew when you meet. The boxes are first raters as far as he could see but he could not see inside the locks and these were the veriest trash. The fellow saw that was the only hole in which he could cheat and gain or rather save a sixpence.

The locks were all useless before we landed in Africa — weak gingerbread things though the maker knew what they were [needed] for — they were even much smaller than the holes made for the keys indicated and put on with a slip of tin thus.

[Two sketches appear in the original]

The slip of tin in some cases was only at the bottom and fell back or moved away from the bolts at once on the smallest pressure — some of the bolt pieces fell back at once and were useless. If Mr B meets the fellow let him have a word of advice but I fear the only word he would need would be a kick.

D. L.

1. Bisa: amale, finger millet (Mr P. Chilekwa).

132. TO JAMES YOUNG
[text badly torn]

[Bangweolo?]

[July? 1868]

I began at 14° to prepare letters for home and mentioned what the Portuguese had said — His ignorant remark was taken as better than six lunar distances — [said] the folly of the alteration becomes apparent when the seventh does not move the spot perceptibly nearer to the imaginary midway. To my gentle remonstrance against this he returned such a [gibe] — other positions were desecrated in the same idiotic way. I am the only traveller that ever gave all his notes and observations for examination by the [geographer of the] council Arrowsmith — Speke preferred burning his to letting them be pryed into by the busybodies who lounge at their rooms in Whitehall.

Arrowsmith and Galton wrote out 'Instructions' for me and say 'considering the large sum they [had voted] £500, they must have copies if not the originals of all my notes. That is because they give ¼ part of my expenses, they must have my private memoranda. [All] who serve me will have a good lump of wages to shew. When I have finished I shall have nothing except empty fame and sit down as a slave and copy my notes for the gaping busybodies of the Council.

I did not study [histrionics] till I was out here and the conclusion I have come to is this — I shall give neither notes, observations nor sketches until after publication. If this displeases Sir Roderick I shall pay back the £500. If the Council cannot guard my observations I can. Only should anything happen to me they would be lost altogether. I wasn't sure my notes on geography to be [commented] on by the Loungers of the society's rooms among whom is Baines who three [?] times confessed to me that he had given away the Government property — offered to pay for the stores he had stolen and begged in the most abject manner to be allowed to remain with the expedition without pay. I had to dismiss him afterwards not so much for his theft as for getting hold of the store book and forging entries [of] the issue of twelve months provisions for six persons in three months to [?] persons. He says that I was kind to him all through but I was led on by my brother, and the Council believes him and in [?] charity makes their rooms his show place [?] they pay no rent for them to Government. I have been thus minute because you gave full half of my expenses. I would have no objections to shew you all my notes if you desired them but to the ignorant loungers I will not shew. I saw that the private notes of my first journey were copied by someone with tracing paper and resolved never to give them again. From the Zambezi I gave every one of my observations in two volumes and they remained for months in Arrowsmith's hands. From this *After* publication is the word for observations and notes not at all. Which but snobs would ask them. The *Instructions* are rich as containing words the meaning of which Galton did not understand — by 'Hydrography' for instance he meant rainfall. 'A survey' [?] £500. It would have taken £5000 to have surveyed Rovuma.

Leaving these follies I am often distressed in thinking of Tom — ought to have been at home to assist and guide him. He was to have two years at Glasgow college and a year or more at a German university or in Germany for French and German and preparation for the civil service examination for India. I did not anticipate being so long. If at home I could speak to Sir Charles Wood. And [?] now I send a note to Sir Bartle Frere who will willingly assist Tom as he did Kirk on my petition. Tom had better be well prepared for examination. I think that some of the ministry would help him on; Lord Palmerston sent [?] gentleman to ask me what he could do for me as he [*1 line illeg.*] It never occurred to [?] I was out here that he meant anything for myself or for my family [?] thought only of my work in Africa in which from private notes I knew he took a warm interest. Prince Albert was our enemy and hand in glove with the Portuguese, but we shall speak of that some other time in [sorrow] that I was not selfish enough, and perhaps that was best. I need not say to you my friend help Tom to get a good education — though he may have chosen another [line the] education is indivisible . . . [*bottom line of page missing*]

. . . me at home as well as my money when I am abroad. I advised you by a letter sent in February 1867 to send £400 to Captain Fraser of Zanzibar [or to] House Fleming & Co of London which I left at interest with Messrs Coutts & Co. I hope you received that [?] I need it for wages and goods. I send for more from Bombay for I give the Arabs, who have been overflowing in their kindness to me agreed return in presents. If I had good but not very expensive guns and good watches they would serve me a deal, but you are not in the way of dealing in the things and Sandy Bryson's things gave no satisfaction. A gold watch bought for a Portuguese cost £2 more than it would have done in London. It lost five minutes daily and the regulator did not affect it. I was mortally ashamed of it for the Portuguese will think I have cheated him. This is a private note and not for Hannan or the Professor. I had written to the Dr but my letter [?]. I shall not stop a moment after I have finished but to finish and perfect the discovery, I am sure you will approve. Your judgement and mine usually coincide and I write to you boldly about Tom and give kind . . . [*bottom line of page missing*]

---

133. TO W.C. OSWELL

Near Lake Bangweolo

8 July 1868

My Dear Friend,

You must take this as all I can give you at present. I began a letter to you to be finished at Ujiji, but an inundation prevented my going that way. A letter to Lord

Clarendon if published will explain what I have been doing. I hope I am not premature in saying that the sources of the Nile arise from 10° to 12 south — in fact where Ptolemy placed them. The Chambeze is like the Chobe 40 to 50 yards broad — but the country is not like that at all, it is full of fast flowing perennial burns — we cross several every day — and crossed the Chambeze in 10° 34′ south. It runs west into Bangweolo — leaving that Lake it changes its name to Luapula — then into Lake Moero. On leaving it the name Lualaba is assumed.

The Chambeze, Luapula and Lualabe receive thirteen streams from 30 to 50 yards broad and always crossed either by bridges or canoes. Three of these I have not seen. Then Lualaba is joined by Lufira, a large river which drains the western side of a great valley in which lie extensive copper mines from which we see bars of copper from 75 lbs to a cwt all over the country. They are made like capital I's. Lufira is fed by five streams. Another line of drainage runs into Lake Liemba — one the Lofu I measured at a ford 294 feet say 100 yards — High and waist deep flowing fast towards the end of the dry season. This has eleven burns from four to fifteen yards broad as its sources. To these four a fifth flowing into Tanganyika must be added. There are twenty-three sources flowing in three lines of drainage.

Now the uncertain part is West of Tanganyika. Some say that large body of water goes past that lake on its west into [*Tanganyika, deleted by D.L.*] Lake Chowambe which I take to be Baker's Lake others that it enters Tanganyika and still goes to Chowambe by a river named Loanda. That is the part I have yet to explore and the everlasting tramp on foot is tiresome.

I have suffered much needless annoyance by two blockheads, the busybodies on the council writing "instructions" for my guidance and demanding all my notes "*copies if not originals*". Because the society pays one fourth part of my expenses I am to sit down as a slave and copy for it the only property I shall have left. If I were to get £2000 after finishing instead of nothing after finishing the £2000 I would not be so annoyed by the snobs asking private memoranda though I think few English gentlemen would have done it. Speke preferred to burn all his notes and observations too to submitting them to the busybodies on the council.

One of them, Arrowsmith, has altered my geographical positions most recklessly and tacked on 200 miles of Lake to the narrowest end of Nyassa. I wish a gentleman had been employed who know[s] the *usus loquendi*[1] of words. Galton, who I supposed penned "Instructions" uses "Hydrogeography" for rainfall. The word is restricted to the survey of coasts or inland waters having a navigable communication with the ocean. I am to make a survey. I preferred to follow Sir Francis Beaufort's advice to Arctic explorers — "Gentlemen remember you go on discovery not on survey" i.e. never spend time on measuring if you can discover. I am to take latitudes every night ! ! !

I hope you are playing with your children insteading of being bothered by idiots.

In looking back to Kolobeng I have but one regret and that is that I did not feel it my duty to play with my children as much as to teach the Bakwains. I worked very hard at that and was tired out at night. Now I have none to play with. So my good friend, play while you may. They will soon be no longer "bairns".[2] My kind love to their mother and them.

David Livingstone.

1. *Latin*: way of speaking.
2. *Scots*: children.

134. TO W.C. OSWELL

October 1869

Manyuema Country,
say 150 miles west of Ujiji.

My Dear Friend Oswell,
I have not the faintest prospect of being able to send a letter for many months to come, but I want to be partially prepared for the time when the bustle of putting up a parcel may arrive. I do not feel it right on those occasions to give my friends a hurried scrawl and when an Arab party is met with going to the coast the headman can scarcely be expected to feed 200 or 300 people for 2 or 3 days merely to let a foreigner pen words which may be against the Arabs in general and himself in particular. I shall try to give you now what may be of permanent interest if as usual it be viewed through the indulgent medium of your friendship. Although it may not have been fully explained in my letter to you from Ujiji you must have learned a little otherwise about what I called the central line of drainage of the great upland valley that begins in 12° South Lat: it goes through Lakes Bangweolo, Moero and another. I saw it as the Lualaba going out of Moero and away North West before I went to Ujiji as soon as I recovered from pneumonia which was worse than ten fevers — that is fevers treated by medicine and not the dirt supplied to Bishop Mackenzie at the Cape — I went fifty miles up Tanganyika [and] southwards to avoid a great mass of high mountains opposite Ujiji on the western shore, and then went Nor West into the Manyuema or cannibal country. I was at a loss about the Lualaba for it was reported to enter a third lake and on coming out it flowed West no one knew whither. Its great size made me fear it was the Congo. Here I found myself in the great bend it makes before turning North or North East into as I as yet only conjecture the Nile. There can be little doubt that such is its destination but I must go down and see — It is very large and were it not so very Burtonesque I would call Tanganyika [and it] Lacustrine rivers — extant specimens of the Lake rivers which abounded in Africa in prehistoric times and along which we often found our smoothest waggon paths — The conviction has slowly crept over my mind that these two great lake rivers are the two arms into which Ptolemy makes

the headwaters collect. As he places the [sources correctly] in Latitude, and gives as correct a view of their division into two headwaters as could be expected from one copying oral information, his authorities [or] probably the informers of his predecessors must have visited these very parts. In fact all that we moderns can fairly claim is the *rediscovery* of what had sunk into oblivion like the circumnavigation of Africa by the Phoenician admiral of one of the pharoahs B C 600. He was not believed because he said that 'in passing around Libya he had the sun on his right hand' — This to us who have passed round the Cape from East to West stamps his tale as genuine. The position of Ptolemy's sources as his two headwaters shew equally reliable geography — The high watershed though only 6000 ft in altitude must be his mountains of the moon, but why he called them Lunae montes[1] I cannot divine — He did not believe in snow causing the inundations though later writers did, nor in the Etesian breezes, but all I recollect of the little I read of his geography makes me speak with great hesitation — Kenia and Kilimanjaro though said to have snow could not be meant — They are so very far from the sources and have no connection whatever with the Nile — The crescentic shape of the watershed could scarcely have given rise to the name or Ptolemy would have given it that shape instead of making them (the Mtns) like cocked hats. The ancients were pretty free with castles, elephants, gorgons, fishes etc. but it was left to Arrowsmith to perch 200 miles of Lake on mountain tops. A map of the Ethiopian gold mines of the time of Sethos II is said by Dr Birch in one volume of the Archaeologia to be the very oldest in the world. Sesostries made maps and distributed them. I should like to see a specimen for all the very old maps I have seen contain a good deal of Ptolemy's idea — He is by no means the first who gave African information. The ancient visitors to these parts were probably traders in ivory, gold, copper, slaves and tamed elephants — I found a tradition among the cannibals that their forefathers tamed and rode elephants — This is not of much worth except when contrasted with the total absence of the idea south of this — to those with whom I have spoken there it seemed as strange as riding on the moon would have been — A plant you mentioned as put on the hedges used in elephant capture and which is dreaded by the animal is here unknown — I never asked you what it was — but I supposed it to be a climbing plant — If I had been able to point it out and enquired its uses I might have revived the tradition — without it no hedge would keep elephants in. It is on record that tamed elephants were brought to I think one of the Greek commanders when out on a foray — He bought some and would have bought more but the owners like wise men preferred to eat them — This has been quoted as conclusive evidence of incapacity for civilization — They preferred four or five tons of good fat meat to three or four brass pots — and so should I. Their wives made clay pots and it was only Greek [brass] that could ask the animals which had cost the labour of several months of the people of a large district to capture to be handed over for their useless wares. I heard a Scotchman quote non-taming of elephants as proof of non capacity for civilization and we Scotch could never tame our own Highlanders. I feel rather at a loss how to speak about poor Speke's discoveries. If I say nothing about them I shall give offence — If I say that Ptolemy in his small Lake 'Coloe' gives a more correct view of Victoria Nyanza than Speke and Grant offence will still more be taken — He affords the best example I know of the eager pursuit of a foregone conclusion — When he discovered Okara on Victoria Nyanza he at once concluded that herein were the sources of the Nile and he would allow no one to question his conclusion — 20,000 square miles of water ! ! ! he would not permit his own mind even to suspect a flaw, and he conjectured a 'backwater' to eke out his little river when he saw that it would not account for the Nile — In V Nyanza lay the sources and no mistake. Mr Baker seems oddly enough to have upheld Speke's conclusion — though the sources are from 500 to 700 miles further up the great Nile valley than what were supposed to be the Fountains by Speke, Grant and Baker. No large river begins in a lake [torn] this one seems to be enormously exaggerated in order to make the rise of the Nile therein feasible — If the Suahili who have lived for years east of Okara be not greatly mistaken, three lakes have been run into one large Victoria Nyanza — Okara has many large islands, and large spaces are covered with 'Tikatika' — a mass of aquatic vegetation in which a grass the Lotus and Duckweed are the chief ingredients. It can be walked over and we were told of its existence at Gumladow. Were it not for the current which bears up miles and miles of confervae[2] and aquatic plants Tanganyika would be largely covered too. I feel a little sorry for Speke's friends but my regret is lessened by remembering that he went out of of his way to say at a geographical meeting that Portuguese crossed the continent before I did and that Dr Roscher discovered the North end of Nyassa before we did the south end. The two slaves on whose journey from Cassange to Tette the Portuguese rest their claims went less by 600 miles than from sea to sea, if back again would have been [an]other 600 or say 1000 in all. And Roscher came near the middle and as Col. Rigby's despatch incontestably proves, two months *after* we discovered the south end — The meeting had not this subject under discussion but in this and other cases he thought that he was imitating the dashing style of Burton, but for whose evil teaching Speke would have exhibited the qualities of an English yeoman. Grant needs no pity. The sources led to his getting a good wife — £2000 a year, and a London house with her, though he never saw them — I have to go down and see where the headwaters join — then finish up by going round outside or south of all the sources — seeing if I can in the way the underground rock excavations and the copper mines of Katanga which have been worked for ages and the malachite is said to be inexhaustible — I don't like to leave my work so that another may 'cut me out' and say he has found sources south of mine — I am dreaming of finding the lost city of Meroe at the confluence of the two head branches — the reality reveals that I have lost nearly all my teeth — that is what the sources have done for me, and while I feel that all my friends will wish me to make a complete work of the exploration I am at times distressed in thinking of my family — a slate fall on Tom's head and injured his eyesight. I ought to have been at home to help him to a situation if he is recovered. I cease writing for the present with my kindest love to Mrs Oswell and her young people.

Affectionately yours
David Livingstone

Margins: There are three lines of drainage Ptolemy knew but two — the water-shed is between 700 and 800 miles long from West to East. Four fountains rise from a mound on it each not ten miles apart from the others. One I name after you by anticipation. It is the source of the Kafue. No water can flow up the bank on south side of the watershed so I give up going round.

Two of the fountains mentioned flow north — two form Zambezi. These are probably the fountains of the Nile mentioned by the secretary of Minerva in the city of Sais to Herodotus and from which half the water flows north to Egypt, and the other half to 'Inner Ethiopia' (I hope to reach there long before this reaches you)

I have been greatly hindered by want of attendants. This is November. Because no envelopes I send this to Agnes. I long sorely to retire.

. *Latin*: Mountains of the Moon.
. Water plants.

135. TO W.C. OSWELL

Manyuema
November 24th 1870

cut a leaf out of my Bombay cheque book to add a little to your letter. I am not quite sure that my packet of letters was destroyed but it contained cheques for goods and other [men] and was sent in June 1869 yet nothing has come in reply — the plunder of goods and the detention of my box for which 15 dollars were paid to carry it to Ujiji either directly or indirectly by the man who went with Burton and Speke and contrived to make them pay so enormously [*torn*] me with grave suspicion — If you have received a bird's eye sketch of the sources having neither longitudes nor latitudes and the Lakes relatively wrong towards each other you will know that the packet has not been destroyed. I sent that even with great reluctance because possible information by means of imperfect sketches which I imagined would serve till my positions were correctly calculated at the Cape.

I sent an imperfect sketch of Nyassa, the Shire & with this understanding. It was given to Arrowsmith either by you or by Mr Bates or was taken away without leave long before I came home — I allowed Kirk to look on my work and mentioned the latitudes and longitudes as I calculated them in a hurried [?]. He sent a copy thereof. The astronomy was done exclusively by myself — Kirk and my brother took time for one only — When I came home I found that my sketch had been thrown aside for Kirk's which is no better than a copy. I asked in your presence if my sketch had not come home long ago. Oh yes, said Bates, 'long ago'. You ordered it to be used but I said 'never mind it'. I saw at a glance that the small geographers meant to exalt Kirk and soon his sketch appeared emblazoned on the walls in large

169

letters 'Dr Kirk's sketch map of Nyassa', a simple falsehood. Well Arrowsmith worked away at my sketches and then at Dr Kirk's though he said it was forty miles wrong and mark you not employed by Mr Murray or me. He was employed after some of my observations came home from the Cape — but he still continued at Kirk's copy till the last batch of obns came from the Cape and he charged Mr Murray and me too because he had to alter work on receiving letter [?] and he saw the remark by the assistant Astronomer Royal Mr Mann "that from the repeated series of chronometric observations given from the coast and from Tette the position of Nyassa could not possibly be 4' wrong and it was very satisfactory to find by the observations made at some of the same spots as on my great journey from Loanda to Quillimane that the first set were generally correct." These were the positions of which Arrowsmith made such havoc and then wrote to Germany and India in self-glorification — It would be unwise in me therefore to let my feelings of affection for you to prevail so far as to place any other observations in your power — Don't blame me — I sent a sketch of Victoria Falls and when I wanted it applied to Mr Bates and was answered not by him but by old Mrs Baines!!! I had actually to tell the old hag that I wanted none of her son's property but my own.

Of the £300 charged by Arrowsmith because he had to alter his work before I came home & before he was employed by Mr Murray or me the unbusiness way of doing work at the rooms [?] Dr Shaw left let me in for at least £200 — The bird's eye view purposely wrong though it may be is better than Arrowsmith's best map or indeed any other — for after all the cooking of accounts by Cooley — the 'erudite' 'the learned' the 'profund' you cannot tell the position of Casembe within a hundred miles.

I am sorry to be in a manner compelled to speak disparagingly of the opinions of my predecessors in exploration, but the discovery of the sources of the Nile was asserted so positively and withal so honestly that explanation is necessary in making a similar claim — Poor Speke's great mistake was the eager pursuit of a foregone conclusion — When he discovered Victoria Nyanza he at once leaped to the conclusion that therein lay the sources of the river of Egypt — 20,000 square miles of water confused with immensity. He would then allow no doubt to arise even in his own mind. As soon however as he and Grant looked to the Nyanza they turned their backs on the *caput Nili*[1] which is 500 miles further up the great Nile valley than the most southerly point of Okara of V Nyanza — When he saw that little river that issues thence would not account for the Nile, instead of conjecturing a backwater whatever that in nature may mean, but for his devotion to his foregone conclusion he would have gone west into the trough of the valley and there not to mention the lower Tanganyika the upper part of which he had already participated in discovering, he must have come to Webb's Lualaba not 80 or 90 yards as his per *White Nile* but never less than 2000 [yards and] often from 4000 to 8000 yards, and then still further West another great Lacustrine river of equally enormous size. The true headwaters of the river of Egypt — I sympathize with all real explorers and admire the splendid achievements of Speke, Grant and Baker. A Dutch lady explorer awakens my admiration greatly though I never saw her. She had provided

vith such wise foresight for both land and water travel and with her steamer
roceeded so far and persevered so nobly in spite of the severest domestic afflic-
ion — the loss of her two aunts by fever — that had she not been honestly enough
f course, assured by Speke and Grant that they had already discovered in Victoria
Nyanza the sources she sought — she must inevitably by boat or overland have
eached the headwaters — I cannot conceive of a lady of so much pluck stopping
hort by Bangweolo. We great He donkeys say exploration was not becoming her
ex — Well considering that at least sixteen hundred years have elapsed since the
ources were formerly visited and Emperors, Kings, Philosophers, all longed to
now the fountains whence flowed the famous river and longed in vain, exploration
oes not seem to have been very becoming the other sex either — She came further
p than Nero Caesar's centurions. What honour has been bestowed. [*incomplete*?]

. *Latin*: head of the Nile.

---

136. TO JAMES YOUNG
[Copy in David Livingstone's handwriting]

am sorry to hear of the Bombay affair but I shall be able to give you something for
have got salary from Govt and a most kind despatch from Lord Clarendon now
las gone from us.

Jjiji 16th Dec 1871

---

137. TO W.C. OSWELL

Tanganyika
6 Jan 1872

.S. I send what was written long ago and will feel obliged if you allow me to use
our name on a fountain from which the Kafue is said to rise. You remember that
Mokantju told us long ago that the Liambai or Upper Zambezi and the Kafue rose
t one place and separated. Kafue is called Lunga Loenge (hard g) — Kafuye and
.efue. It seems to be one of four fountains rising very near to each other and I am
resumptuous enough to think that these are the ancient fountains of Herodotus of
hich we all read in boyhood and rejected as we became wise in natural

171

philosophy. Mokantju was right and I have heard of the earthen mound at which
the four fountains rise so often — and know pretty [well] the four rivers they form
that I venture to say I wait another year to rediscover them. When we heard
Mokantju's tale we were about 350 miles from the mound but Liambai whose
fountain I call Palmerston's and Kafue whose fountain I call Oswell's do most
certainly flow into Inner Ethiopia. The other two Bartle Frere's flows as R Lufira
into lake Kamolondo and Young's (I have been obliged to knight him to distinguish
him from the gunner, as Sir Paraffin Young) goes through Lake Lincoln into
Lualaba and North to Egypt. They thus [form] the central line of drainage, Webb's
Lualaba into Petherick's branch.

But I have been sorely bamboozled by my friend Kirk handing over the matter of
supplies and men to a rich Banian slave trader who naturally wishes me anything
but success and all his slaves came deeply imbued with the idea that they were not
to follow but force me back. Then they ran riot on my goods for 16 months instead
of three and ultimately sold off all for slaves and ivory to themselves. Well with a
loss of these which amounted to £500 or £600 — It is not very edifying to hear that
Kirk again took half £1000 sent by Govt to Ludha again and he sent slaves again
who feasted three and a half months on the mainland and then have spent from Oc
1870 to this in jollification on the way and I am obliged to go back and catch the
fellows 150 miles off. The slaves forced me back 400 or 500, then I go
150 + 150 — and back from this to my work. It cannot be less than some 1800
miles, all because Kirk finds it easy to say to a polite Banian take this two day
before he came. I feel very sore at this "companion" of Livingstone, Sir Roderick
calls him.

Love to Mrs Oswell and children
David Livingston

I add a few words more to thank you for your kind attention to Oswell who I am
glad to hear has taken to the medical profession of his own accord. He must work
his own way up. Sir Bartle Frere says that he will do what he can for him if he goes
to India but I don't know enough of the medical profession there to direct him.
suspect he must go to America, for all advance in our country is through interest
and if I am removed he will have none.

Did I dream that Baker had all his guns taken from him by niggers of whom he
speaks so contemptuously — I could not pitch into even slaves without being
certain of finding them all gone through the first night afterwards, but he thrashed
them and the Arabs "cussing and swearing promis[?]" and they carried him
meekly while I have to tramp every step I go. His statements seem to me insuffer-
able brag — and Sir Roderick, good man, takes his tales as all sound gospel. He
apportioned the sources of the Nile to him who turned 600 miles short of them and
to Speke, Finlay, and Arrowsmith because they dreamed about them. He does not
himself know the structure and economy of the watershed or the way that this
structure correlates the economy of the great Lakes and Lake rivers in the
phenomena of the Nile. I am thinking of offering a prize of beef steaks and stout for

an explanation from Sir Roderick's pets before I divulge it all to you. I say the structure of the watershed is exactly what you and I found at [?Ntlotle] in the Kalahari, that and the enormous Lacustrine rivers and Lakes are the means which kind Providence arranged to regulate the [grand] old Nile, but this is for you and Mrs O. Willie must never say he shot an antelope at 600 yards with the weapons that Baker used — "drop" is the word. He meant 600 feet surely but I earnestly hope he will stop the Nile slave trade, and I shall forgive all his bosh of discovering sources.

138. TO JAMES YOUNG

[15 March—July 1872]

P.S. Agnes told me that you had most kindly written to Kirk to draw on Bombay from you on my behalf. He said that he had drawn the money he squandered on Ludha but did not say how much. I suppose it must have been £600. When asked for the account of expenditure he merely said that he had handed off all that remained to the Search and Relief Expedition and Stanley had got £500 from it. He evidently does not wish me to see how very liberal he has been with my money. The slaves for instance got three times full freemen's pay — that is from 25$ to 30$ a year is the highest freeman's pay. He gave them $30 down and said that their pay of 60$ a year began on that day. Now I shall need at least £600 to pay all my men on reaching Zanzibar and I told Kirk so and said if you had authorized him to draw for me to do before I reached the coast. I did not then know that he had mounted his high horse and publicly declared before five responsible witnesses that "he would do nothing for Dr Livingstone because he would only get insulted for I believed niggers". It was his own belief in niggers and Ludha that caused all the losses and that I believed them is only an inference drawn by himself. I never said that I either believed or disbelieved them. I reported what they said when they swore and stood to it that Dr Kirk ordered them not to follow me but force me back. What I say to you confidentially I did not say to him that his private self invitation to share in the honours — his public advice for me to retire and leave the work to others published to Banians and all others.

It is uncalled for mixing up his name with a discovery by a false description gathered from Arab [?]. His lazy indifference in handing off large sums of money without a single precaution. His eagerness in writing home "I have sent off men and goods to Dr L". All Dr L's wants have been supplied — The Government having by private offer and their plan of relief their sending out a fresh expedition points to one already and eager to share in the honour and then his gratuitously taking blame to himself for the assertions of the slave and without the smallest proof saying that I believed them — make our friend's conduct look grave. His public refusal to do anything for me seems to me that he will not draw even if authorized to do so. I mention the matter to you and hope that you will not feel that I am bothering you as I suppose many do. I do not think like my sisters and son Bill bickering that they have a perfect right to my money, that I have any right to yours. I give you a hint and leave the matter entirely in your hands.

I would not say so much but that I have good reason to believe that Dr Kirk will *not* do what you may be expecting him to perform.

David Livingstone

APPENDIX
I

'PUBLIC MEETING,
AUGUST 12TH,
IMPORTATION OF COOLIE & CHINESE
LABORORS.' [sic]

From *The Natal Times*, 24 October 1851 page 3

Mr Millar, being called to the chair, urged the consideration of the present question, upon the Meeting, by contrasting the present insignificance of the colony, with the importance, which it was capable of attaining under an ample supply of labour.

Mr A. W. Evans rose to move the following Resolution.

That it is impossible to rely upon the Kafir population of this Colony, for a permanent and effective supply of labour, and that, successfully to raise tropical production, it is absolutely necessary to introduce free foreign labour.

At a season of the year when food was least plentiful among the Kafirs, labour was not to be obtained. The cry was universal, from the Quathlambas to the sea coast — the scarcity of labour — many were abandoning agricultural pursuits, many leaving the country in disgust, exports we had none, — the colony was being drained of its specie by thousands of pounds. In England, he had thought with others, that, by proper management, Kafir labour might be made available: he had lived to see the fallacy of this opinion: in proof of the aversion of African savages to work, he had known men in the Bechuana country to go into the fields, and dig up roots for their subsistence, rather than take a spade in hand to earn 6d a day for food: the Kafirs might indeed be relocated in less formidable bodies, but, with the first year's harvest would return the aversion of the men to that labour, which the custom of generations has imposed solely upon their women. It was doubtful, if even the suppression of Polygamy would force these people to work: but if coercion (and what coercion was so stringent as that of depriving a man of his wife?) were desirable, Government was in no present position to try it with the Kafirs: it was from consciousness of this kind that the collection of the tax had, last year, been suspended. The analogous case of the Frontier Kafirs, the increasing boldness of language and demeanour, evincing a growing consciousness of their strength, in our own natives, these were proofs of the light in which, refugees as they are, these people would regard themselves, and that they never could be forced to work until decimated in years of war. Such a war, to be successful on our side, would require a large military force, it would need every colonist to be a soldier, every English artizan to use his gun as readily, as he now handles his tools. English emigrants were inexperienced in the culture of tropical productions, the work was unsuited to their habits, most of them aspired to be proprietors themselves, and the rest would expect £20 per annum wages, with rations worth another £20; this was obtained in Australia but we could not afford to pay so much for labour here. Juvenile paupers had been tried at the Cape, until their maintenance in due subordination had been denounced as quasi slavery at Exeter Hall. The desired impetus to the prosperity of

175

Appendix I

this colony must come from India or China: with the cheap labour which could be obtained from those countries, we should import an additional safe-guard to the colony. Coolies and Chinese, from their differences in habits and religion, would not only form no coalition with Kafirs, but, like the Fingoes on the Frontier, be our best bodyguard against them. He could positively state, that merchants in England were only waiting for a satisfactory settlement of the labour question to invest thousands of pounds in Natal. The soil must be made to yield a return for this money: We might be well off in the town, but the town could never make the country, agriculture alone could create permanent wealth, solid prosperity; to promote it we had capital enough, capital in the colony, capital coming here, labour and labour only was required.

Mr Searle seconded the resolution.

Transcribed from a photocopy of the original, kindly provided by the Librarian, Killie Campbell Africana Library, University of Natal, Durban

APPENDIX
II

First impressions of
the Victoria Falls, November 1855. Notes used
in writing *Missionary Travels*

. . and prolonged from the left end of the tunnel[1] through 30 miles of hills and the pathway being 100 ft down from the bed of the river instead of what it is with the lips of the fissure from 30 to 100 ft apart then fancy the Thames leaping bodily into the gulph and forced there to change its direction and flow from the right to the left bank then rush boiling and roaring through the hills and he may have some idea of what takes place at this. The most wonderful sight I had seen in Africa.

In looking down into the fizzure [*sic*] on the right of the island one sees nothing but a dense white cloud which at the time we visited the spot had two bright rainbows on it. (The sun was on the meridian and the declination equal to the latitude of the place.) From this cloud rushed up a great jet of vapour exactly like steam and it mounted up 200 or 300 feet high. There it changed its hue to that of dark smoke and came back in a constant shower which soon wet us to the skin. This shower falls chiefly on the opposite side of the fizzure and a hedge of evergreen trees whose leaves are always wet. From their roots a number of little rills run back into the gulph but as they flow down the column of vapour licks them up clean off the rock and away they mount again. They are constantly running down but never reach the bottom.

On the left of the island we see the water a white boiling mass moving away to the prolongation of the fizzure near the left bank of the river.

The fizzure is said by the Makololo to be much deeper farther to the Eastwards but there is one part at which the walls are so sloping that people accustomed to it can go down by descending in a sitting position. The Makololo on one occasion pursuing some fugitive Batoka saw them [? unable] to stop themselves in the descent literally dashed to pieces at the bottom. They behold the stream like a white cord at the bottom and so far down (probably 300 feet) that they became giddy and were fain to go away holding on to the ground.

Now though the edge of the lip over which the river falls does not shew wearing more than three feet and there is no appearance of the opposite wall being worn out at the bottom in the parts exposed to view, yet it is probable that where it has flowed along beyond the falls the sides of the fizzure must have given way and the parts out of sight may be broader than the white cord' on the surface.

A piece of rock has fallen off at a spot on the left of the island and juts out from the water below and from it I judged the distance which the water falls to be about one hundred feet. The walls of this gigantic crack are perpendicular and composed of one homogenous mass of rock. The edge of that side over which the water falls is worn off two or three feet and pieces have fallen off so as to give it a somewhat serrated appearance. That over which the water does not fall is quite straight except at the left corner where a [?] appears and a piece seems inclined to fall off. Upon the whole it is nearly in that state in which it was left at the period of its formation. The rock is dark brown in colour except about 10 feet from the bottom where it is discoloured by the annual rise of the water to that or a greater height.

177

Appendix II

On the left side of the island we have a good view of the mass of water which causes one of the columns of vapour to ascend, as it leaps quite clear off the rick and forms a thick unbroken fleece all the way to the bottom. The whiteness gave the idea of snow, a sight I had not seen for many a day. As it broke [? wild] pieces of water all rushing on in the same direction each gave off several rays of foam exactly as bits of steel when burned in oxygen gas give off rays of sparks.

The snow white sheet seemed like myriads of small comets rushing on in one direction each of which gave off (as they are represented on paper) from its nucleus streams of foam. I never saw the appearance referred to noticed elsewhere. It seemed like the effect of the mass of water leaping at once clear of the rock and but slowly broken up into foam.

I have mentioned that we saw five columns of vapour ascending from this strange abyss. They are evidently formed by the compression suffered by the force of the water's own fall into an unyielding wedge shaped space. The enormous mass coming down in constant flow must [? batter] that already there in somewhat the same way that air is compressed in piston to produce fire. Of the five columns two on the right and one on the left of the island were the largest — And the streams which formed them seemed each to exceed in size the falls of Clyde at Stonybyres when that river is in flood. This was the period of low water in the Leambye but as far as I could guess there was a flow of five or six hundred yards of water which at the edge seemed at least three feet deep. I write in the hope that others more capable of judging distances than myself will visit this scene and I state simply the impressions made on my mind at the time.

I thought and do still think that the river above the falls to be one thousand yards broad — I am a poor judge of distances on water for I shewed a naval friend what I supposed to be 400 yards of the bay of Loanda and to my surprise he pronounced it to be 900. I tried to measure the Leambye with a strong thread, the only line I had in my possession, but when the men went two or three hundred yards they got into conversation and did not hear our shouting when the line became entangled. By going on they broke it and being carried away down the stream was lost on a snag. In vain I tried to bring to my recollection the way I had been taught to measure a river by taking an angle with the sextant — that I once knew it and that it was easy was all I could recall and that only increased my vexation. I however measured the river further down by another plan — and then I found that the Portuguese had measured it at Tete and found it a little over one thousand yards. At the falls it is a broad [?] more so, so whoever may come after me will not I trust find reason to say I have indulged in exaggeration.

Presented to the Livingstone Museum by Miss A Thom

1. The Rotherhithe Tunnel under the Thames in London.

APPENDIX
III

Note concerning some lakes in East Africa and the Zambezi and Shire rivers, by the Viscount de Sá da Bandeira from *Boletim Official do Governo da Provincia de Moçambique*, 26 October 1861

Since the famous Doctor Livingstone, after travelling along the Shire River in Zambezia, has written that the Portuguese had not navigated this river, we were curious enough to examine certain books with the aim of verifying what knowledge they contained about the lakes and rivers mentioned above, before the recent journeys were undertaken; and have made the following extracts from them:

1.

In the work called *Ethiopia Oriental*, written by Father João dos Santos, who for eleven years from 1586 to 1597 was a missionary in the Portuguese lands of East Africa, one reads the following: 'Facing Sena, on the other side of the (Zambezi) River, stands a very large and high mountain called Shire (at present known as Mt Murrumballa), which can be seen from twenty leagues. This mountain is most fertile, and is inhabited entirely by Blacks, both on the heights and in the valleys. From there come most of the supplies used in Sena, such as rice, corn, potatoes, bananas, and chickens.

'There are many excellent springs of water, not only in the valleys but also on the heights. At its foot flows a large and beautiful river, which joins the Zambezi ten leagues below Sena, and along it travel the Blacks and the (European and Asian) inhabitants of Sena, and their trade from one place to another passes along it.'

2.

Father Manoel Godinho, in his journey by land from India to Portugal in 1663, speaks of Lake Zachef' and says it is fifteen leagues wide, of unknown length, and he adds the following:

'According to a map I saw, made by a Portuguese who travelled for many years in the kingdoms of Monomotapa, Manica, Butua, and others in the lands of the Blacks, this lake is not far from Zimlaré, that is to say the court of Mezura or Marabia. From it flows the river Aruri (probably Luangwa), which joins the Zambezi above our fort at Tete; and also the river Shire, which after flowing through many lands and finally those in the neighbourhood, joins the river Cuama (Zambezi) below Sena. That such a lake exists was told me not only by the Blacks but also by the Portuguese, who had already reached it by travelling along the above mentioned river.' (Godinho, *Viagem*, 2a edicão, p. 199, Gamito, *O Muata Cazembe*, p. 48)

179

Gamito, speaking of the lake or river 'Nhanja Mucuro', that is 'Big Nhanja', says that according to information he received, it is extraordinarily wide, so that in crossing it in canoe one is forced to spend two nights on islands, with which it is scattered, arriving on the opposite shore on the afternoon of the third day. By his calculations this is a distance no exceeding nine leagues. The lake has a strong current towards the east and many of the islands are inhabited.

He says that there is another river called 'Nhanja Pangono', that is 'Small Nhanja', by the Blacks. (*Muata Cazembe*, p. 48).

According to the same traveller, the principal lakes seen by the expedition commanded by Major Monteiro were Lake Luena and Lake Mofô, both in the dominions of Mwat Cazembe. He says that the latter has no apparent current, that its length from south to north is impossible to see, and that by what he heard, the lake does not empty its waters into an other; and that into it, from the south, flow the river Canenjué and the river Lunda, both having much water, and perhaps others (*Muata Cazembe*, p. 447).

He says too that the city of Lunda where the Mwata Cazembe resides is situated in a wid plain on the eastern side of the large lake Mofô, here more than four leagues wide and with low and muddy shores; the lake abounds in fish, crocodiles, otters, and zovôs, some sort o amphibious animal said to be the size of a goat (p. 243).

Doctor Livingstone, referring to a conversation which he had at Tete, in the house of th Commandant of the town in February 1856, says on page 640 of his *Travels*, (1857 edition) 'One of the gentlemen present, Senhor Candido, had visited a lake 45 days to the NNW o Tete, which is probably the Lake Maravi of geographers, as in going thither they pas through a people of that name. The inhabitants of the southern coast are named Shiva; thos on the north, Mujao; and they call the lake Nyanja or Nyanje, which simply means a larg river. A high mountain stands in the middle of it, called Murómbo, or Morumbola, which i inhabited by people who have much cattle. He stated that he crossed the Nyanja at a narrow part, and was 36 hours in the passage. The canoes were punted the whole way, and if we tak the rate about two miles per hour, it may be sixty or seventy miles in breadth. The country al round was composed of level plains covered with grass, and indeed in going thither the travelled seven or eight days without wood, and cooked their food with grass and stalks o native corn alone. The people sold their cattle at a very cheap rate. From the southern extremity of the lake, two rivers issue forth; one, named after itself, the Nyanja, which passe into the sea on the east coast under another name; and the Shire, which flows into th Zambezi, a little below Senna. The Shire is named Shirwa at its point of departure from th lake, and Senhor Candido was informed, when there, that the lake was simply an expansio of the river Nyanja, which comes from the north and encircles the mountain Murómbo, th meaning of which is junction or union, in reference to the water having been parted at it northern extremity, and united again at its southern . . . The Portuguese are unable t navigate the Shire up to the lake Nyanja because of great abundance of a water-plant whicl requires no soil, with the name 'alfacinya' (*Pistia stratiotes*) from its resemblance to a lettuc . . . Senhor Candido holds the office of judge in all the disputes of the natives and knows thei language perfectly.'

This Senhor Candido is probably the same individual who on 3 June 1831 was encountere by the expedition led by Major Monteiro in the Prazo Soche on the left of the Zambezi whei he was en route from Tete to Cazembe.

The said individual, who was at Luane, the residence of the prazo mentioned, and who wa called Candido José da Costa Cardozo, held the office of Grand Captain of the Lands of th

Appendix III

Crown, and Special Judge in the cases of Blacks (pleas and other questiones between Blacks). (V. Gamito, *Muata Cazembe*, p. 7).

4.

River Zambeze (i.e. Chambeshi): Gamito (in *Muata Cazembe*, p. 191) says that on 9 October 1831 Major Monteiro's expedition arrived on the left bank of the River Zambeze, the name of which is spelled Zambeze in the manuscript; he observes that this river runs towards the west, that it was eighty fathoms wide and five fathoms deep between the banks, that its bed was of stone, well populated with oysters which were excellent to eat, and rooted in 3½ palms of water, flowing at high speed, not less than that of the Zambezi, in some places nine miles per hour: that it is a river of considerable size with few crossing points even in the month of October when because of the summer weather all rivers are low.

He says moreover on page 387 that on the return from Kazembe the expedition arrived on the right bank of the same river on 18 July 1832; that it was 100 fathoms wide and the depth between the banks at the crossing point was eight fathoms, but that the speed of the current over a bed of smooth and slimy rock made the crossing impracticable. He says that on the journey to Kazembe he crossed the river further to the east, but was unable to calculate the distance between the two points; that he supposed it not to be small, because the terrain where the first crossing was made was mountainous, while in the second place the river runs through a vast plain; that at the first point there was an abundance of oysters in the river, and at the second, no trace of them.

On page 447 he says that of all the rivers the expedition crossed between Tete and Lunda, Kazembe's capital, the river Chambeze is the one with the greatest volume of water, and appears to be the one which could be made navigable with the greatest ease: and that he considers it possible that it joins the Zambezi.

Let us record here that the merchants who have penetrated the interior from Angola mention as the source of the river of Sena, which enters the sea below Quelimane, a large river whose name they write in the following manner — Diambege — Liambege — Riambege.

5.

Reflecting on what Major Gamito says about Lake Mofô, and considering that its position is to the north west of Moiro-Acheisto and Masavamba, whose latitudes have been determined by Doctor Lacerda, it appears that the position of this large lake should be a few days' journey from Lake Tanganyika, which was visited by Captains Burton and Speke between 1857 and 1858.

6.

It seems that the following conclusions may be drawn from these extracts:
(1) That the river Shire had been navigated by the Portuguese in the 16th and 17th centuries
(2) That the large lake called 'Nhanja Mucuro', i.e. 'great water', which exists in the country of the Maravi, had already been visited by them in the 17th century
(3) That they have made maps in which the said lake, the Shire, and other rivers have been marked
(4) That in their journeys between Tete and Kazembe they had often crossed the river Zambeze or Coambeze

Appendix III

(5) That Candido da Costa Cardozo had been at the place where the river Shire emerges from the lake, Shire having there the name Shirwa

(6) That Doctor Livingstone in visiting the Shire and fixing certain points by astronomical observations and making a description of the country, added to the geographical knowledge there was about this part of Zambezia.

Lisbon, January 1861

Sá da Bandeira
*Literal translation by Evitar from photocopy
of original kindly supplied by the librarian,
Arquivo Histórica Ultramarino, Lisbon*

APPENDIX
IV

19 November 1862

Note explanatory of a map published by the Viscount de Sá da Bandeira and sent to different European governments as a "New Portuguese map"

This map when taken in connection with a paper by His Excellency in the Mosambique *Boletim* of 26 October 1861 affords pretty plain proof that the Viscount's geographical knowledge is by no means equal to his patriotism.

The first glaring error that meets the eye is the large portion of the Zambesi which is made to flow down to Quillimane. The Zambesi and the River Quillimane are as distinct as the Thames and the Humber — only, at high flows when the delta of the Zambesi is nearly submerged several meandering creeks therein are filled the two rivers may be said to inasculate by that of Mazaro, but except at the [time] specified not a drop of Zambesi water goes to Quillimane. The Zambesi at the Mazaro creek is about a mile wide and merchandise is carried a good day's journey overland from it to the river of Quillimane; the true mouths of the Zambesi are never used except by the English. The explanation came out here that the Quillimane river has been represented to foreigners as the main outlet of the Zambesi is that the slave traders thereby keep the attention of English cruisers directed to Quillimane, while they use the real mouths about sixty miles to the south for the undisturbed shipment of slaves.

It is interesting to observe how eagerly the good Viscount contends for the extensive geographical knowledge of his famous ancestors while with all ancient and modern Portuguese researches at his command he propagates the grievous error at the very outset of his map.

In marking districts H. E. seems to have been guided by the "inevitable destiny" doctrine that makes Cuba a part of the United States, for his lines only indicate the size the district may attain some 200 years hence. The right bank of the Zambesi for instance belongs to a tribe of Zulus called Landeens and to think the Portuguese now pay a heavy annual tribute for the small strips of land they occupy around Senna and Shupanga — the annual rent for the former is 3,200 yards of calico with a quantity of beads and brass wire for the latter about half that amount — they would not occupy Shupanga at all but for the huge trees found there which are converted into large canoes. During our late operations at Shupanga we saw the Zulus come repeatedly to lift the tribute and they never demanded anything from the English. They conquered all the country southwards of the river and down to Sofala and beyond that — and even drove the people of Senna from their fort to the islands in the Zambesi. Previously to that a Portuguese gentleman came into possession of the large estate called Chiringoma by marriage with a chieftain's daughter — but the government of Portugal partitioned it on the ground, it is said, that it was wrong for a subject to possess land larger than Portugal. The partitioning never come actually into effect in Africa because in the meantime the estate had changed masters by Landeen conquest.

All these portions of this so called "new map" west of the Loangwa [Arrongua] are copied without acknowledgement from my own earlier map. Only the spelling of the names are

183

Appendix IV

altered to suit Portuguese othography. All previous Portuguese maps and even that of Mr Bowditch exhibit the source of the Zambesi where we discovered the Victoria Falls. No better test of ignorance of the interior could have been furnished than absence of notice of a scene which for size and strangeness is unequalled in the world. When two hundred miles distant the question by a native chief, "Have you any such in your land?" showed the impression they had made on the savage mind. Supposing he had meant an ordinary waterfall my reply in the affirmative elicited the rejoinder, "What causes the smoke?" That smoke or vapour I have since seen with the naked eye at 21° of latitude distant. If an overland route ever existed, knowledge of this marvellous scene must have come to the coast.

The Shire too is copied from our own earlier tracing, but the Viscount at his desk has made it arise by numerous streams lying parallel with Lake Nyassa, while we have since seen that it flows out of the lake in one goodly stream alone and having trudged over every foot of the right bank can declare that the Shire receives not a single feeder where he has put so many branches.

As the paper in the *Boletim* asserts that I have done injustice to all the Portuguese of the 16th and 17th century, quotations are given from Padre J. dos Santos's work *Ethiopia Oriental* to the effect that "dwellers in Senna and the caffres of the mountain Morambala navigated and carried on commerce from the one place to the other" that is, that early intercourse between a village and mountain about 20 miles apart about 300 miles of river from the Lake Nyassa prove that this lake was visited by Portuguese in the C16 and C17th. Captain Wilson went 200 miles of river distant from Nyassa, no one or at least no Englishman would quote him as its discoverer in the 19th century.

The other quotation is that Padre Manoel Godinho while on the overland route from India in 1663 had heard something of Lake Zachaf and had seen a map drawn by a Portuguese who had travelled about in the Kingdoms of Monomotapa, Manica, Butua, etc. which showed the lake to be near the "court of Mezura, or Marabia" all which parts being south of the Zambesi and at least 400 miles of river distant from Nyassa, we read as if it were said, someone in going to England in 1663 saw a map drawn by somebody else who having travelled about many years in Algarve (see map of Portugal) had inserted a lake situated as if at Salamanca in Spain but placed in the Alentejo, and to this incongruous assertion hearsay navigation up to the lake Nyassa is appended in innocent ignorance that Murchison's Cataracts render that impossible.

Maps have even been drawn by men who never left their easy chairs and such map makers have even ventured to dispute the existence of rivers etc. with those who have actually seen and sailed on them. Native information has also been appropriated by Europeans and possibly some more reliable geographical knowledge may have been buried in the Portuguese archives but it may be presumed that only when the interior and lakes were visited by Englishmen and the results given to the world that the discoveries can be said to be made at all. With every disposition to appreciate the Viscount's patriotism when he sat down and made a jumble of the world wide discoveries with previous Portuguese errors — then modestly called the result a "new Portuguese map" I think he might have been better employed.

Instead of copying the Portuguese discoveries in the background as the Viscount asserts, his paper proves that I did spontaneously publish all I could to the credit of Senhor Candido of Tette before any Portuguese even mentioned his name. He was met by their own traveller Gamitto and supplies of bread given by him are duly acknowledged, but not a hint given that he had discovered a lake till years afterwards when I found him at Tette. Well he has always pointed to the lake he is supposed to have visited as north of N N W of Tette and asserts that the shores of his lake were covered with large cabbage shaped trees with soft stems, that the

Appendix IV

canoe was punted or poled all the way across a narrow part near its southern end which required 36 hours to cross — that there is a mountain called Murombua or Murombola in the lake — the name meaning union from the waters in their flow from it to the south parting at one end and uniting at the other. Now Lake Nyassa on the contrary is NE or NNE of Tette. There are no soft stemmed cabbage shaped trees on its shores and at a mile from the bank no bottom can be found with a line of 100 fathoms or 600 ft. The broadest part would not occupy 36 hours in a canoe and the narrower portion requires only six hours. There is no mountain — not even a hill in Lake Nyassa. The islands are small uninhabited rocky islets with one exception and that is called Zumora and no current whatever can be detected — the line even at 100 fathoms hung perpendicularly from the boat. As Gammitto mentions from hearsay "a strong current towards the east" in the lake and Candidio sees two rivers flow from the south end; the only way that his statement can be received is on the supposition that there is another lake somewhere between Tette and Tanganyika as yet not visited by anyone capable of laying down its position.

The names Shirwa Nyanza — Nyanja or Nyassa, prove nothing as they are applied to marshes, rivers and lakes: Shirwa Manga the great foreign water means the sea — Nyanja (great or little) is applied to the Shire — to the marshes in it, to Lake Nyassa to Shirwa and to Tanganyika.

When one of the most enlightened and benevolent of the Portuguese statesmen propagates the grave geographical errors referred to above and strives to appropriate large sections of territory over which his nation has no control, it may be inferred that the recent offers of farms to foreign emigrants though probably quite sincere are not to be trusted. Their own subjects have to pay heavy annual tribute to the native owners of the land. The map is indicated as a claim on the land interior by those who have never done aught on the coast but convert the country into a slave preserve to enrich needy governors — and as their right to the coastline is allowed only on condition of putting down the slave trade between Cape Delgado and Delagoa Bay this new claim ought certainly to be disallowed.

David Livingstone

APPENDIX
V

MEMORANDUM
21 December 1864

Memorandum of an agreement between David Livingstone now residing at Newstead Abbey in the county of Nottingham in England LLD of the one part and Charles Livingstone Esq now about to proceed to Fernando Po as her Britannic Majesty's consul of the other part as follows . . .

1st. The book which is now in the course of preparation being a Narrative of the late expedition to the Zambezi and of the Discovery of Lakes Shirwa and Nyassa on the East Coast of Africa shall be forthwith published both in England and America under such title and in such manner as the said David Livingstone or his publisher Mr Murray, shall determine, and so as best to secure the copyright in the said book in the respective countries.

2nd. Should it be deemed expedient, with the view of securing the copyright in the United States, the said Charles Livingstone shall be at liberty to publish the said book within the United States, in his sole name or take any other steps in that country for securing the sole benefit to himself thereof in that country.

3rd. In whose name or names soever, or in what manner soever the said book shall be published whether in the United Kingdom or in the United States, the profits of such publication shall be divided as follows viz:

All profits whatsoever arising therefrom, derived from the sale thereof in the United Kingdom or any of the colonies or dependencies thereof shall belong to and be the absolute and exclusive property of the said David Livingstone his Executors administrators and assigns.

And all profits whatsoever arising therefrom derived from the sale thereof in the United States of North America shall belong to and be the absolute and exclusive property of the said Charles Livingstone his Executors administrators and assigns.

4th. The said Charles Livingstone hereby authorises the said David Livingstone to do all acts and execute all deeds or other instruments for giving full effect to the foregoing agreement. As witness our hands this twenty first day of December in the year one thousand eight hundred and sixty four.

Witness to the signature of the said David Livingstone	John Kirk Witness Horace Waller Witness
Witness to the signature of the said Charles Livingstone	[Illegible]
	DAVID LIVINGSTONE CHARLES LIVINGSTONE

(Braithwaite)

Index and Notes

This index contains brief biographical, topographical and other material about many entries, where readers may wish for information about them. Proper names unindexed include members of correspondents' families where these occur in greetings, enquiries as to health etc., London street names, and the authors and titles listed on page 40.

Abbreviations and usage:
CMS: Church Missionary Society
LJ: D Livingstone, *Last Journals*
LMS: London Missionary Society
NEZT: D Livingstone, *Narrative of an Expedition to the Zambezi and its Tributaries*
DLCD: *David Livingstone: a Catalogue of Documents* and *Supplement*
RGS: Royal Geographical Society
UMCA: Universities Mission to Central Africa
South Africa: The Republic of South Africa
south Africa: broadly speaking, Africa south of the Zambezi and Kunene rivers

Aberdeen, 142
Abyssinia: i.e. modern Ethiopia 154
Academy, The: Royal Academy of Arts, London 89, 129
Adams, Charles Francis (1807—86): US Minister to Britain 1861—8 100, 116
Adams, Henry Gardiner (1811—81): perpetrator of the unauthorized *Dr Livingstone: His life and adventures in the interior of South Africa* (London 1857) 95
Ajawa: 71, 72, 74, 84
Albert, Prince (1819-61): Prince of Saxe—Coburg, consort of Queen Victoria 164
Alentejo: province of Portugal 184
Alexandria, Egypt 150
Algarve: province of Portugal 184
Algoa Bay: site of the present Port Elizabeth, South Africa 1, 6, 13
Ali: servant of Richard Thornton q.v. 83
Almeida: see Tavares
Almorah. Uttar Pradesh. India: 27
Amoda: 91, 147, 149
American Board of Commissioners for Foreign Missions: founded 1812; in 1835 sent Rev

Aldin Grout, Rev G Champion and Dr Newton Adams to establish missions in Natal, south Africa 27
American Minister: see Adams, C F
Angas, George Frederick (1822—86): artist in New Zealand, Australia and the Cape Colony, where he specialized in African portraits
Anjuan: island in Comoros group, east Africa 69, 71, 123, 137, 160
Anthropological: i.e. the Royal Anthropological Society, founded 1863 by Richard Burton q.v. 136
Apothecaries Hall: Hall of the Honourable Company of Apothecaries (est. ca 1600), Virginia Street, Glasgow 161
Arabs: Livingstone's expression for Afro—Arabs of east Africa, particularly Zanzibar. See Swahili 63, 87, 155, 156, 158, 160, 164, 166, 172, 173
Archaeologia: Society of Antiquaries publication 167
Arctic explorers: 165
Argyll, George Douglas Campbell, 8th Duke of

Index

(1823—1900): Liberal politician, British Secretary for India 1868—74. Owner of the island of Ulva, Scotland, where the Livingstone family originated 50, 89, 91

Arnot, David Jr (1821—1894): Son of Scottish immigrant to Cape and his Khoi wife. Qualified in law at Cape Town, practised at Colesberg. Defender of Griqua rights 36

Arrowsmith, John (1790—1875): Cartographer, drew the maps for Livingstone's books 98, 135, 151, 163, 165, 167, 169, 170, 172

Ashton, William (1817—97): member of LMS, mainly at Kuruman; printer 25, 28

Aston: 132

Atheneum The: British weekly literary review 152

Augustine, Saint (d. AD 604): Missionary sent from Rome by Pope Gregory to convert the English to Christianity 117

Babisa: see Bisa

Babolebotla: probably the Ila people of today's Zambia, who used to remove the upper front teeth 36

Bagamoyo: 148

Baines, John Thomas (1820—75): Official artist and storekeeper to Zambezi Expedition, dismissed by Livingstone 1859. Author of The Victoria Falls, Zambezi River, Sketched on the Spot (London 1865). Recent writer, eg Ransford, Jeal, have sought to exonerate Baines from Livingstone's charges against him. 43, 46, 51, 54, 62, 64, 66, 67, 133, 163

Baines, Mrs: Mother of Baines, J T q.v. 170

Baker, Sir Samuel White (1821—93): British explorer of the Nile, 1860s 142, 153, 165, 168, 170, 172, 173

Baker's Lake: i.e. Lake Sese Seko, between Uganda and Zaire; named Lake Albert (after Queen Victoria's husband) by Samuel Baker 1864 165

Bakhatla: the Tswana Kgathla clan 22, 23, 32

Bakwain: the Tswana Kwena clan 4, 23, 27, 39, 165

Baloochees: Baluchi, a people of today's Pakistan 103, 137

Bamangwato: the Tswana Ngwato clan 25

Bamapela: Tswana clan 24, 35

Bambarre: 145

Bandeira: see Sa

Bangweolo: Bangweulu, large lake and swampy area in northern Zambia 145, 146, 147, 148, 165, 166, 171

Banian: Indian merchant class 172, 173

Bann: see Russell J S 55, 57, 58

Bartholomew, Hugh: Manager of Glasgow City and Surburban Gas Works 162

Basiamang: half-brother of Chief Sechele q.v. 38

Basileka: Tswana clan 24

Bates, Henry Walter (1825—92): Naturalist, Assistant Secretary, Royal Geographical Society 1864—92 169, 170

Bath, western England: 89, 91, 95, 102, 103, 116

Batoka: 177

Beaufort, Sir Francis (1774—1857): Royal Navy Admiral, naval hydro-geographer, inventor of Beaufort Scale for wind velocity 165

Bechuana: see Tswana

Bedingfeld, Norman Bernard (1824—94): Royal Navy Commander whom Livingstone first met at Luanda, Angola; second-in-command Zambezi Expedition, but after disagreements, soon resigned 43, 45, 54, 55, 56, 57, 65, 67, 79, 102

Beke, Dr Charles Tilstone (1800—74): British explorer and Biblical scholar 153, 154

Belchior do Nascimento: Portuguese of Tete district, alleged by Livingstone to be a slave trader 140

Bemba: a people of northern Zambia. Livingstone also uses the name for Bangweule q.v. 159, 161

Benguella: port in Angola 87

Bennett, Dr (later Sir) James Risdon (1809—91): Son of Secretary of LMS, befriended Livingstone in London 1840. Elected President of Royal College of Physicians 1876 29—33

Bethelsdorp: LMS station near Port Elizabeth; founded 1803 15

Birch, Dr Samuel (1813—85): Egyptologist, from 1861, Keeper of Egyptian and Oriental antiquities, British Museum 167

'Birkenhead Jews': i.e. Macgregor Laird's shipyard at Birkenhead, England: 'jews' used derogatorily 65

Birt, Richard (1810—92): LMS missionary in south Africa 22

Bisa: a people of northern Zambia, established extensive trading links between the interior and Lake Malawi 62, 63, 159, 160

Biscay, Bay of: 7

Blue Book: British parliamentary report: in this instance, Accounts and Papers 25, Slave Trade Session, 1862 (DLCD) 126

Boatlanama: place, with waterhole, in Botswana 37, 38

Beers: Dutch-speaking settler farmers (boers) of south Africa. 1, 4, 35, 38 Today called Afrikaners. See Swahili

Boma: a tropical nut-bearing tree, Vitex 112, 124

Bombay: Administrative and trade centre, west coast of India 97, 128, 150, 153 164, 169

Bombay affair: probably reference to the crash of the Agra Bank, resulting in the loss of Livingstone's savings from the sale of Lady Nyassa 44, 117, 147, 171, 173

Index

Borneo: large island between Malaysia and Indonesia 100

Boston: main city of Massachusetts, USA 95, 109

Botletle: river flowing out of Lake Ngami q.v.

Bowditch, Thomas Edwards (1791—1824): British geographer, author of *An Account of the Discoveries of the Portuguese in the Interior of Angola and Moçambique from Original Manuscripts'* (London 1824) 184

Braithwaite, Joseph Bevan (1818—1905): lawyer and friend of Livingstone 65, 66, 132, 186

Brentwood, Essex, England: 35

Bridges, Charles (1794—1869): Evangelical Christian writer on religious matters 10

Brighton, in southern England 73

Bristol, in western England 116

British Association for the Advancement of Science: founded 1831 89, 91, 95

British Museum: national museum of Great Britain, London 84

Brock: 53

Brown, Dr: 139

Brown, Richard: Manager of James Young's q.v. chemical works at Bathgate, Scotland (Strathclyde archives) 132, 136

Brown, Dr Thomas Joseph D.D. (1798—1880): British Roman Catholic Bishop 139

Brownlow and Oliver: London tailors 68

Bryson, Dr Alexander (Sandy) (1802—69): Friend of Livingstone from college days in Glasgow. Director General of Royal Navy Medical Department, Honorary Physician to Queen Victoria 103, 142, 164

Buchanan, Dr Andrew (1798—1878): Professor of Medicine, Glasgow University; teacher of Livingstone and later one of his trustees 13, 73, 85, 158, 162

Burnbank Road: address of the Livingstone house in Hamilton 135

Burrup, Mrs Elizabeth Mary: wife (widow) of H de W Burrup q.v. 76, 78

Burrup, Henry de Wint (1831—62): Anglican minister, member of UMCA. Died of fever 44, 72

Burton, Mrs: wife of R Burton q.v. 97, 98

Burton, Richard Francis (Sir) (1821—90): British adventurer, explorer, writer and translator. Visited Lake Tanganyika 1856 with J H Speke q.v. 60, 66, 79, 91, 98, 136, 150, 168, 169, 182

Bushmen: derogatory term for the San of south Africa 67

Butua: 179, 184

Buxton: health resort, Derbyshire, England 137

Byron, George Gordon, Lord (1788—1824): renowned poet, at one time owner of Newstead Abbey q.v. 97

Cabango: village in eastern Angola 41

Cabora (or Cahora) Basa: deep gorge on lower Zambezi, first visited by Livingstone 9 November 1858. Now largely flooded by hydro-electric dam 43, 46, 58, 67, 108, 109, 114, 120

Caffre: i.e. Kaffer (unbeliever), derogatory Arab and later European term for Blacks, especially in east and south Africa 38, 175, 176

Caffre War: Livingstone is referring to the 'War of the Axe' 1846—7, the seventh war in the European conquest of the eastern Cape colony 28, 38, 80, 175

Cain: Son of Adam and Eve 36

Calliatte, Revd: French Protestant minister 144, 150

Candido: 180, 184

Cannibals: 166

Cardoso, Candido Jose da Costa (1800?—90): Trader and landowner at Tete. Military captain-in-chief, Tete 1861. May have visited Lake Malawi 1846 180, 182

Casai: i.e. Kasai, river of Angola and Zaire 41

Cassange: inland town, Angola 168

Cataracts (Murchison's): series of waterfalls on Shire River, Malawi; named after Murchison q.v. by Livingstone. Now site of hydroelectric scheme 46, 70, 184

Cazembe: see Kazembe

Ceylon: Sri Lanka 53

Chambers Journal of Popular Literature, Science and Arts: published Edinburgh, Scotland 66

Chambeshi: large river in northern Zambia 147, 148, 165

Chambeze: see Chambeshi

Chapman, James (1831—72): Trader, author of *Travels in the Interior of South Africa* (London 1868) 67

Chewa: People and language of Malawi and eastern Zambia (where the language is called Nyanja). Successors to the Maravi q.v. Livingstone's Manganja were Chewa. 'Formerly all the Manganja were united under their great chief Undi, whose empire extended from Lake Shirwa to the river Loangwa' (NEZT, p. 198). The present Chief Undi resides in Zambia 91, 98, 136, 150, 168, 169, 182

Chibisa: Chief on Shire river, at the village now called Chikwawa 50

Chihombo: river in eastern Angola 41

Chikape river in eastern Angola 41

Chipeta: a people of south-western Malawi 159

Chitambo: 146, 148

Chobe: river in northern Botswana, visited by Livingstone and W. C Oswell q.v. 1851 4, 59, 165

Chonuane: i.e. Chonwane, Livingstone's second mission station, 75 km north of Zeerust,

Index

Transvaal, South Africa 1, 4, 24, 25, 30, 31, 32

Chowambe: Livingstone's hearsay lake in southern Zaire; perhaps Lake Upemba 165

Chuma: see Susi 91, 147, 148, 149

Church Missionary Society (CMS): Anglican evangelical missionary society founded 1799 14

Clarendon, George William Frederick Villiers, 4th Earl of (1800—70): British Foreign Secretary three times between 1853 and 1870. 165, 171

Clementine: i.e. Clementino da Souza, ivory trader, merchant, mayor of Tete. From Goa, India, initially a slave trader 82

Clifton: suburb of Bristol, in western England 96

Clowes, William: London printer 107, 110, 111, 113, 114, 119, 121, 130

Clyde: river, Scotland: 177

Colesberg: trading centre, Cape Province, South Africa 24, 36

Collingwood, Cuthbert: 50

Coloe: see Ptolemy

Comoros: Indian Ocean islands: 137, 145, 147

Committee of House of Commons: in this instance, Select Committee of Africa (West Coast), 18 May 1865 134

Confederates: the southern states during the American civil war 1861—5

Congo: alternative name for River Zaire 166

Constantinople: i.e. Istanbul, Turkey 113

Conyingham, George Lennox (?—1866): Chief Clerk, Foreign Office, London 80, 143, 150, 153, 154

Cooke, Robert (?—1891): Cousin and partner of John Murray q.v. 120, 121, 125, 127, 128, 129, 130, 138

Cooley, William Desborough (?—1885): Author of *Inner Africa Laid Open: in an attempt to trace the chief lines of communication across the continent south of the Equator* (London 1852) 170

Crawford, John: 152, 153

Cotton, Colonel: W C Oswell's q.v. uncle 97

Coutts: (i) Coutts and Co, British banking house 61, 84, 159, 160, 164
(ii) Miss Angela Georgina Burdett-Coutts (1814—1906): wealthy heiress and shareholder in Coutts & Co; gave large sums for humanitarian and Christian causes, endowed Anglican bishopric of Cape Town, supported Livingstone 100, 142

Cruz, Antonio Jose da Cruz Coimbra: alleged slave merchant of Quelimane 72

Crystal Palace: large iron-framed glass building erected in London to house a great imperial exhibition, 1851 129

Cuba: Republic in Caribbean; in the mid-19th century there was a movement in Cuba to escape Spanish colonial status by joining the USA 112, 183

Cumming, Roualeyn George Gordon (1820—66): hunter and trader in south Africa; relation of Duke of Argyll 25, 26, 153

Dakanamoio: island in the Shire river, near Chief Chibisa's town 62

Daubeny, Rev Dr Charles Giles Bridle (1795—1867): Professor of Chemistry, Oxford; President of British Association 1856 138

Daucan: 134

Davies, Captain: Commander of *Hetty Ellen* q.v. 76

Dean of Ely: see Harvey, Goodwin

Delagoa Bay: site of Maputo, Moçambique 25, 185

Delgado, Cape: cape on north Moçambique coast 185

Dechmont: mansion and estate in Livingstone parish, Linlithgow, Scotland; Dechmont House near Dechmont Hill (686 ft/221 m high) 49 (Mr B Cunningham)

Denman, Capt.: 56—7

Dent: watchmaker in Strand, London 65, 84

Diario de Lisboa: i.e. Lisbon Daily, Portuguese newspaper 116

Dickie,: 38

Dollars: the Austrian Maria Theresa dollar was a popular means of exchange in east African trade during Livingstone's time there, but as there was extensive trade between the US and Zanzibar at the period, the US currency may be that referred to

Dos Santos, Rev Joao: Portuguese missionary in east Africa, 1586—97 179, 184

Dover: port on English Channel 142

Downs, The: safe, anchorage for shipping, near Deal, Kent, England 7

Dreux: town in department of Eure et Loire, France 144, 150

Duncan, Captain: Commander of *Pearl* q.v. 67

Duprat, Alfredo (1816—81): Portuguese consul and representative on anti-slavery commission, Cape Town. See Frere, G 77

Dutch lady/ladies: see Tinné

Ecclesiastes 116

Edwards, Rogers (1795—1876): artisan (carpenter) member of LMS at Kuruman. Founded Mabotsa q.v. with Livingstone 14, 15, 17

Edwards, Samuel (Samo) Howard (1827—1922): trader, son of R Edwards q.v. 67

Egypt: 5, 92, 150, 169, 170

Elwin, Whitwell (1816—1900): editor *Quarterly Review* 1853—6; editor standard edition of Pope q.v. 130

Emma, Queen: wife of King Kamehameda of Hawaii 142, 143

Index

Index

192

Index

Kamolondo: Livingstone's imaginary lake in southern Zaire 172

Kariba: 43

Kasungu: 44, 47

Katanga: the present Shaba province, southern Zaire 87, 148, 168

Kazembe: also Cazembe, Mwata Kazembe: 63, 87, 145, 147

Kebra Basa: see Cabora Basa

Keelwa: see Kilwa

Kendal, north-west England 73, 86, 97, 98

Kenia: i.e. Mt Kenya, snowcapped mountain in east Africa (Kenya) 167

Kilimanjaro: snowcapped mountain in east Africa (Tanzania) 167

Kilwa: in Livingstone's time, important 'Arab' trading centre, southern Tanzania

Kirk, Dr (later Sir) John (1832—1922): medical officer and botanist on Zambezi Expedition; later British consul, Zanzibar 43, 44, 45, 46, 52, 60, 62, 63, 66, 68, 71, 75, 82, 83, 84, 87, 92, 96, 100, 115, 116, 135, 142, 143, 145, 147, 148, 151, 152, 153

Kirk's Range: the western Malawi escarpment, named by Livingstone after John Kirk—'The whole chain from the west of the cataracts up to the north end of the Lake' (NEZT p. 491)

Knapp's Technology: i.e. *Chemical Technology or Chemistry Applied to the Arts and to Manufactures* by Dr F Knapp, Giessen University, Germany. 3 volumes, illustrated (London and Paris 1848) 65

Kolobeng: Livingstone's third and last mission station; in Botswana some 400 km north-east of Kuruman, 16 km south east of Molepolole 1, 4, 25, 38, 164

Kololo: Sotho-speaking conquerors of the upper Zambezi region (see Sebitwane). Sekeletu, the second Kololo king, provided Livingstone with over a hundred porters, and financial support, for the transcontinental journeys. Many of Livingstone's 'Makololo' were from subject peoples, e.g. Tonga. The name Kololo is said to derive from 'Makollo', the clan to which Sebitwane's favourite wife, Setloutlou, belonged. See D. F. Ellenberger, *History of the Basuto* (London 1912). 4, 45, 46, 52, 58, 60, 64, 87, 100, 106, 107, 117, 130, 177

Kongone: an entry to the Zambezi delta from the sea 62, 69, 70, 78, 107, 112

Koribelo, Botswana 37

Krapf, Dr Johan Ludwig (1810—81): German born member of the Church Missionary Society; active in east Africa. His *Vocabulary of Six East African Languages* (Tubingen 1850) was reviewed in the *Church Missionary Intelligencer*, October 1850. Krapf in his introduction writes 'One common language lies at the bottom of all the idioms spoken from the Equator to the Cape of Good Hope'. 36

Kroomen: men of the Kru people, renowned seafarers of Sierra Leone and Liberia, west Africa. Twelve were recruited at Freetown, Sierra Leone, for the Zambezi Expedition. The nicknames only, of only six, are known—Tom Will, Tom Peter, Tom Toby, Tom Coffee, Black Will, Tom Jumbo 54, 58, 66

Kuruman: LMS station in today's northern Cape Province, South Africa; founded by Robert Moffat q.v. 1821 beside Kuruman river, near Lithako q.v., capital of the Tswana Tlhaping clan, 1, 3, 4, 13, 15, 17, 19, 22, 25, 32, 35, 41, 45, 46, 86, 91

Lacerda, Jose de: published refutation of Livingstone's claims, *Diario de Lisboa*, 15, 17, 19 December 1864; English translation issued as pamphlet *A Reply to Dr Livingstone's Accusations and Misrepresentations* (Edward Stanford, London 1865) 140, 151

Lady Nyassa: Ship which Livingstone had built at his own expense (£5589 5s 9d) for use as trading and anti-slavery patrols on Lake Malawi. *Lady Nyassa* was built by Tod & Macgregor at their shipyard, Meadowside, near Glasgow (Mr W Cunningham) 44, 46, 47, 48, 60, 75, 84, 85, 86, 88, 91, 93, 121, 123, 133, 142, 151

Lady of the Lake: i.e. *Lady Nyassa*

Laird, Macgregor (1808—61): shipbuilder and merchant of Birkenhead, England. Member of 1832—4 Niger Expedition. Built Zambezi Expedition paddle steamer *MaRobert* q.v. 61, 63, 64, 67

Landeens: Landins was the name given by the Portuguese to the Nguni who migrated from the south into Moçambique, starting about 1821 183

Lattakoo: see Lithako

Layard, Sir Austin Henry (1817—94): British Under Secretary of State for Foreign Affairs 1861—6, best remembered as an Assyriologist and excavator of the ancient city of Nineveh 77, 79, 114, 154

Leambye: see Liambai

Lechulathebe: i.e. Tswana chief, Letsholathebe, near Lake Ngami 100

Lemue, Jean Louis Prosper (1804—70): French Protestant missionary at Motito, near Kuruman. Officiated at Livingstone's wedding to Mary Moffat 144

Lepinole: river, Botswana: 24

Liambai: local name for Upper Zambezi 171, 172

Libya: ancient name for Africa 167

Index

Liemba: local name for Lake Tanganyika 144, 165
Limefield: residence of James Young q.v. 86, 96, 138, 139, 158
Limehouse: dockland district of London 73
Lincoln, Abraham (1809—65): President of the USA during the Civil War 1861—5, in the course of which he abolished slavery; assassinated 1865 130
Lincoln's Inn: London establishment of lawyers' chambers 66
Lincoln, Lake: 172
Lindi: port in northern Moçambique 72
Linyanti: Kololo capital, on Chobe river 119
Lithako: capital of the Tswana Tlhaping clan. The Kuruman mission was established near (New) Lithako, to which site the Tlhaping had moved from (Old) Lithako, 80 km to the north, 1820 35
Litubarula: Capital of Chief Sechele q.v. 39
Livingstone (the name was spelled Livingston by members of the family until about 1857)
(i) Neil (1788—1856): father of David; tea-seller of Blantyre, Scotland 3, 5, 34
(ii) Agnes (née Hunter) (1782—1865): tailor's daughter, wife of Neil 3, 34, 78, 79, 117, 131, 138, 139, 149, 174
(iii) John (1811—99): brother of David; emigrated to Canada; became a farmer in Ontario 66, 68
(iv) David (see below)
(v) Janet (1818—95): David's sister, unmarried 174
(vi) Charles (1821—73): David's brother; married Harriet Ingraham q.v. whom he met while both were students at Oberlin q.v.; appointed British consul, Fernando Po after Zambezi Expedition; died of fever 33, 34, 35, 43, 45, 46, 57, 61, 63, 66, 75, 92, 95—6, 98—9, 109, 135, 141, 169, 186
(vii) Agnes (1823—95): David's sister, unmarried 138, 140
Livingstone, David (1813—73): (ii) Mary (1821—62): wife, 1845, of David; daughter of Robert Moffat q.v. 1, 4, 24, 26, 31, 37, 43, 44, 45, 46, 53, 68
(iii) Robert Moffat (1847—64): eldest son; died as soldier in American Civil War 1, 4, 25, 39, 49, 73, 74, 75, 85, 86, 88, 91, 92, 96, 97, 100, 102, 116, 137, 144, 153
(iv) Agnes (1847—1912): eldest daughter; married Alexander Low Bruce, Scottish businessman, 1875 4, 91, 95—104 *passim*, 112, 122, 123, 124, 129—44 *passim*, 153, 173
(v) Thomas Steele (Tom) (1849—76): second son; died while working in business house in Egypt 1, 4, 85, 86, 91, 92, 99—102 *passim*, 122, 137, 139, 162, 164, 168
(vi) Elizabeth (1850): died a few weeks after birth 1, 4

(vii) William Oswell (1851—92): third son; became a physician and practised in Trinidad 1, 4, 43, 45, 73, 91, 92, 96, 97, 116, 137, 139, 148, 162, 174
(viii) Anna Mary (1858—1939): second surviving daughter; married 1881 Frank Wilson, missionary Sierra Leone, Zambia. Their son Hubert Francis Wilson was for a time a missionary in Zambia and Malawi. 1, 46, 74, 92, 97, 138, 140
Loanda: i.e.
(i) Luanda, capital of Angola 1, 4, 98, 125, 170, 178
(ii) Livingstone's supposed river flowing north out of Lake Tanganyika; probably today's Rwanda 165
Loangwa: i.e. Luangwa, river in eastern Zambia 47, 165, 167, 183
Loapula: i.e. Luapula, large river flowing north in Lake Mweru 159, 165
Lobemba: i.e. Lubemba, country of the Bemba q.v. 161
Lochearnhead: village in Perthshire, Scotland 160
Lochfine: i.e. Loch Fyne, 25 km north-west of Glasgow 160
Lofu: river in northern Zambia 165
Logagen: large cave, used as place of refuge, near Kolobeng 38
Logagen: place in Botswana 38
London Missionary Society (LMS): founded 1795, the LMS was non-denominational but largely Congregationist 1, 2, 3
Luabo: channel in Zambezi delta 76
Lualaba: river flowing north from Lake Mweru, a headwater of the Zaire 145, 147, 148, 166, 170, 172
Ludha, Damji: Indian merchant, Zanzibar 172, 173
Lufira: river of southern Zaire 165
Lunda river: 180, 182
Lunga Loenge: Livingstone's name for the Lunga river, north-western Zambia 171
Lynx: motor ship of Royal Navy 64
Lyra: ship of Royal Navy 68

Mabotsa: mission station founded by Livingstone and R Edwards, near Mafikeng, South Africa 1, 4, 22, 76, 33
Mabrucki: servant of Richard Thornton q.v. 83
Macabe, Joseph (1816—65): hunter and trader in south Africa 37
Mackenzie, Charles Frederick (1825—62): Anglican clergyman, consecrated bishop to lead UMCA; died of fever 44, 46, 71, 72, 74, 80, 86, 88, 91, 107, 110, 117, 122, 127, 157, 160, 166
Mackintosh, James (1766—1879): inventor of waterproof sheeting
Maclear, Sir Thomas (1794—1879): official

Index

British Astronometer at Royal Observatory, Cape Town 85

Maclehose, James (1811—85): Glasgow bookseller and publisher 28

Macmillan: i.e. Macmillan's Magazine (London and Cambridge); published Waller's q.v. 'Scenes in an African Slave Reserve' April 1865, the magazine published by Alexander Macmillan (1818—96) 112, 120

Macrundle Shaw & Co: Glasgow merchants 61, 87

Madeira: Portuguese island in the Atlantic off north west Africa 7

Magomero: 44, 46, 47, 155, 157

Maimeloe: 37

MaJames: see MaRobert

Makololo: see Kololo

Makusa: river north of Quelimane 77

Malagasse: Madagascar?

Malmesbury, James Howard Harris, 3rd Earl of (1807—89): Conservative politician, British Foreign Secretary 1852, 1858—53, 79

Malta 150

Mamire: uncle of Sekeletu q.v. 99

Mamochisane: i.e. Mamotsisane, daughter of Sebitwane q.v.; she was his legitimate successor when he died, but abdicated in favour of Sekeletu q.v. 99

Manganja: see Chewa

Manica: ancient kingdom bestriding present Moçambique—Zimbabwe border 179, 184

'Manifest Destiny': statement of the aspirations of the nineteenth century USA — 'Our manifest destiny is to overspread the continent allotted by Providence for the free develop-ment of our yearly multiplying millions', wrote John O'Sullivan (1813—95) in US Magazine and Democratic Review, July—August 1845. O'Sullivan predicted the expansion of the US until its boundaries embraced the whole north American continent, and Cuba.

Manoel: Moçambique settler 83

Mann, William (1817—73): Assistant astronomer, Royal Observatory, Cape Town 170

Mantatees: the people of Manthatitsi, (female) Sotho chief, set on the move by the Mfecane 36

Manyuema: i.e. Manyema, part of eastern Zaire 145, 147, 166

Maravi: ancient kingdom of western Malawi, after which the country is now named; see Chewa 60, 179, 180, 184

Mariano: alleged slave trader on Shire river 87

MaRobert: the Zambezi Expedition paddle steamer; named after Mary Livingstone (called Ma Robert — mother of Robert—by the Tswana). Livingstone often referred to James Young's q.v. wife as Ma James 45, 46,

53, 56, 66, 68, 70, 71, 109

Marsauceux: village in department of Eure et Loire, France 144

Marseilles: port in southern France 92, 142

Marshall, Miss: 23

Masego: 'Kololo' who worked for Livingstone on Zambezi Expedition 128

Mashue: place in Botswana 31

Mashuwe: town in Botswana?

Masiko: 'Kololo' on Zambezi Expedition who stayed behind when the others returned to the Upper Zambezi 128

Massangano: port in northern Moçambique 72

Massilia: steamship on which Livingstone sailed at beginning of last journey 150

Mataka: Yao chief of Upper Rovuma area, Moçambique 157

May, Commander John Daniel: Royal Navy Assistant Surveyor, Cape Town 69

Mazaro: settlement near mouth of Zambezi 53, 67, 68

Mazitu: Bisa word 'those who come from nowhere' applied to Nguni invaders of Malawi/Zambia in 19th century (Mr E Chuma) 47, 160

Mboma: a Chewa village 124

Mebalwe: Tswana Christian convert and teacher from Kuruman. He came to Livingstone's rescue when, in 1844, the latter was attacked by a lion, which broke his right upper arm. The break, which Livingstone set himself, did not knit true, and made it impossible for him to shoot from his right shoulder. The distorted bone gave confirmation at the post mortem in London that the body brought from central Africa by Livingstone's servant's 1874 was in fact his. A cast of the bone is on display at the Scottish National Memorial to David Livingstone, Blantyre 30, 39

Medlycott, Mervyn B.: Royal Navy Officer 65

Meller, Charles James (1836—69): surgeon to Zambezi Expedition 1861—2 82, 83, 84, 98

Meroe: ancient city and iron-making civiliza-tion near junction of Blue and White Nile 168

Milton: John Murray's q.v. editor 12, 25, 126, 127, 129

Minerva: see Herodotus 169

Misericordia Hospital: at Rio de Janeiro 9

Moatize: coalfield near Tete 51

Mezura or Maralia: ancient kingdom, north-eastern Zimbabwe; perhaps Mazowe

Moçambique: (i) Portuguese fortified town on coast north of Zambezi, until 1897, capital of (ii) Moçambique, the Portuguese colony and today's republic 1, 5, 46, 63, 76, 77, 87, 88, 154

Moffat, John Smith (1835—1918): fourth son of Robert Moffat q.v. Livingstone paid him to establish a mission to the Ndebele 1859—65.

195

Index

Index

Index

Index

Sekwebu had a nervous breakdown, was put in irons by British seamen, and committed suicide by drowning 5, 48

Selwyn, George Augustus (1809—78): Anglican Primate of New Zealand, later Bishop of Lichfield, England; author of *A Verbal Analysis of the Holy Bible, intended to facilitate the translation of the Scriptures into Foreign Languages* (1855) 66

Sena: town on Lower Zambezi 54, 80, 179, 180, 181, 183, 184

Senna: see Sena

Sepoy: Indian soldier in British Indian army 160

Serinan: place in Bostwana 37

Sesheke: town (in Zambia) on Upper Zambezi 47, 107, 118

Sesostris: i.e. the Pharaoh Senosret; according to Herodotus (Two, 103) led an army across Africa, conquering all in his path 167

Sethos II: a Pharaoh 167

Seward, Dr: 152

Shaw, Dr H.Norton (d 1868): Acting Secretary RGS 1854—63 170

Shelley, Edward: big game hunter whom Livingstone met in south Africa 141

Shire: river flowing from Lake Malawi to join the Zambezi 43, 44, 46, 47, 57, 58, 59, 63, 64, 65, 68, 78, 85, 107, 158, 169, 179, 180, 181, 184

Shirwa: lake in southern Malawi 43, 46, 47, 59, 60, 122, 182, 186

Shokotsa: place in Botswana 37

Shoshong: town in Botswana?

Shupanga: settlement on Lower Zambezi 44, 59, 135, 183

Sicard, Tito Augusto de Araujo (1818?—64): Portuguese military commander at Tete 62, 84, 87, 119, 126

Sichuana: see Tswana

Siddim: 'And the vale of Siddim was full of slime pits; and the Kings of Sodom and Gomorrah fled, and fell there' (Genesis, 14, 10) 65

Sinamane: 107

Smith, Sir Henry ('Harry') George Wakelyn (1778—1860), Governor of Cape Colony 1847—52 37

Snow, John: London bookseller 39—40

Soares, Joao da Costa: Moçambique merchant 76

Sofala: ancient Arab trading port, south of Zambezi mouth, Moçambique 183

South Seas/seaism: Christian missionaries were active, and generally successful, in the islands of the south Pacific, e.g. Fiji, Tonga, Hawaii. see Emma 16, 31, 86

Spaulding, Rev Justin: American missionary in Rio de Janeiro, Brazil; perhaps related to Levi Spaulding (1791—1873), pioneer

American Congregationalist missionary in India 11, 12

Speke, John Hanning (1827—64): British adventurer and hunter; visited Lake Tanganyika 1856 with Burton q.v.; visited and named Lake Victoria, declaring it the source of the Nile, a claim disputed by both Burton and Livingstone 66, 89, 91, 141, 152, 163, 165, 167—22 *passim*, 182

Stanford, Edward: London publisher. See Lacerda 140

Stanley, Edward Henry Lord, 15th Earl of Derby (1826—73): British Foreign Secretary 1866—8 136

Stanley, Henry Morton (1841—1904): British-American journalists, author of *How I Found Livingstone* (London 1872) 145, 146, 148, 173

Starkey, S: London hatmaker; made Livingstone's consular caps, and also one for Sekeletu q.v. 68

Steele, Colonel (later Sir) Thomas Montague (1820—90): British army officer in India; met Livingstone while hunting in south Africa 4, 25, 97, 108

Stenhouse, Dr John (1809—80): Chemist, inventor of waterproof fabric and shoes, patented 1861 134, 160, 161

Stephens, Henry Eusely (d 1913): Royal navy captain HMS *Frolic* which took Livingstone and Sekwebu to Mauritius, 1856 48

Steward: 121

Stewart, James (1831—1903): Free Church of Scotland minister; with Zambezi Expedition 1862—3 investigating possibility of founding a mission; in 1876—8, helped establish Livingstonia, Malawi 75

Stonybyres: Clyde, Scotland: 178

Storey's Hotel: in London 127

Sultan of Zanzibar: i.e. Seyid Majid Said (d 1870); became Sultan 1856. See Zanzibar: 93, 148, 153, 157, 158

Suahili: see Swahili

Suez: isthmus and city in Egypt from which the Suez Canal (completed 1869) takes its name 92

Sunley, William (d 1886): Sugar planter, Anjuan q.v. British Consul, Comores 1851—65; said by Livingstone to use slave labour 112 (NEZT p.427)

Susi: 91, 147, 148, 149

Sutherland, Elizabeth Gordon, Dowager Duchess of (1806—68): wealthy British aristocrat and landowner (Sutherland county, Scotland from which she expelled peasants in favour of sheep); campaigner against slavery, friend of Harriet Beecher Stowe q.v. 143, 152

Swahili: Language of east Africa; started as pidgin/creole of Arabic and local (Bantu) tongues; now one of the most widely used languages of Africa and official medium of